6th Photonics Meeting 2023

AA003711

Journal of Physics: Conference Series Volume 2627

Kuala Terengganu, Malaysia
10-11 June 2023

Editors:

Nur Farhanah Zulkipli
Anas Abdul Latiff
Zulzilawati Jusoh
Cheng Xiau San
Zuraidah Harith

ISBN: 978-1-7138-8520-7
ISSN: 1742-6588

Printed from e-media with permission by:

Curran Associates, Inc.
57 Morehouse Lane
Red Hook, NY 12571

Some format issues inherent in the e-media version may also appear in this print version.

This work is licensed under a Creative Commons Attribution 3.0 International Licence.
Licence details: http://creativecommons.org/licenses/by/3.0/.

No changes have been made to the content of these proceedings. There may be changes to pagination and minor adjustments for aesthetics.

Printed with permission by Curran Associates, Inc. (2025)

For permission requests, please contact the Institute of Physics at the address below.

Institute of Physics
Dirac House, Temple Back
Bristol BS1 6BE UK

Phone: 44 1 17 929 7481
Fax: 44 1 17 920 0979

techtracking@iop.org

Additional copies of this publication are available from:

Curran Associates, Inc.
57 Morehouse Lane
Red Hook, NY 12571 USA
Phone: 845-758-0400
Fax: 845-758-2633
Email: curran@proceedings.com
Web: www.proceedings.com

TABLE OF CONTENTS

photonics meeting 2023
gathering . conference . workshop

6th Photonics Meeting 2023 (PM23)

10 – 11 June 2023, Duyong Marina & Resort, Kuala Terengganu, Malaysia

The Photonics Meeting is an annual event organized by the Optical Society of Malaysia (OSM) in celebrating UNESCO International Day of Light. This year's 6th Photonics Meeting 2023 (PM23) was held for two days from 10th to 11th of June 2023. This PM23 is a high-impact conference that was constructed to build a collaboration and joint effort with Universiti Teknologi MARA (UiTM). The purpose of PM23 is to gather experts in Lasers, Optics and Photonics field to contribute in delivering ideas and experiences to all participants involved.

PM23 has attracted 18 contributions of proceeding that cover various topics in the field of Optics and Photonics. During two days conference, PM23 was honored with presentations from five keynote speakers and two invited speakers from universities in the ASEAN region.

Thank you ABEX Sdn Bhd for your valuable sponsorship of the PM23. Your support made a significant impact, and we greatly appreciate your commitment to our event's success. PM23 would like to extend its great appreciation to all the committee members who have work diligently to ensure the smooth deliverance of this scientific event.

All submission has been thoroughly reviewed by the experts in the respective field and presented in this volume of Journal of Physics: Conference Series.

Editors

Nur Farhanah Zulkipli (Tunku Abdul Rahman University of Management and Technology)
Anas Abdul Latiff (UTeM)
Zulzilawati Jusoh (UiTM)
Cheng Xiau San (UOW Malaysia)
Zuraidah Harith (INTI International University)

Content from this work may be used under the terms of the Creative Commons Attribution 3.0 licence. Any further distribution of this work must maintain attribution to the author(s) and the title of the work, journal citation and DOI.
Published under licence by IOP Publishing Ltd

Keynote Speaker

Prof. Ir. Dr. Sulaiman Wadi Harun (Universiti Malaya)
Prof. Dr. Moh. Yasin (Airlangga University)
Dr. Muhammad Imran Mustafa Abdul Kudus (Universiti Malaya)
Ir. Dr. Anas Abdul Latiff (UTeM)

Invited Speaker

Assoc Prof. Dr. Abdel Baset M. A. Ibrahim (UiTM)
Assoc. Prof. Ir. Ts. Dr. Azura Hamzah (MJIIT-UTM)

Organizing Committee

Chairwoman
Assoc. Prof. Ir. Dr. Zulzilawati Jusoh (UiTM)

Chairwoman II
Dr. Arni Munira Markom (UiTM)

Secretary
Ir. Dr. Nani Fadzlina Naim (UiTM)

Treasurer
Ir. Dr Suhaila Subahir (UiTM)

Secretariat
Ts. Dr. Norizan Ahmed (UiTM)
Naimah Mat Isa (UiTM)
Nurul Athirah Mohamad Abdul Ghafar (UiTM)

Protocol & Program Coordinator
Dr Aziati Husna Awang (UiTM)
Dr. Afiq Arif Aminuddin Jafry (Taylor's University)
Norhaslinda Zull Pakkal (UiTM)
Syazwani Mohd Zokri (UiTM)
Nur Atikah Salahudin (UiTM)

Technical Program
Dr. Nur Farhanah Zulkipli (Tunku Abdul Rahman University of Management and Technology)
Dr. Ahmad Haziq Aiman Rosol (Malaysia-Japan International Institute of Technology - UTM)
Dr. Rafidah Rosman (UiTM)

Publication
Ir. Dr. Anas Abdul Latiff (Universiti Teknikal Malaysia Melaka)
Dr. Zuraidah Harith (INTI International University)
Dr. Cheng Xiau San (UOW Malaysia)

Program Book
Syila Izawana (UiTM)
Nuraiza Ismail (UiTM)

Publicity and Marketing
Hasnorhafiza Husni (UiTM)
Ts. Dr. Mohd Talmizie Amron (UiTM)
Dr Baktiar Musa (UiTM)

Designer & Website Developer
Dr. Ahmad Razif Muhammad (Universiti Kebangsaan Malaysia)
Firdaus Jamal Rashid (Universiti Kebangsaan Malaysia)
Muhamad Hakim Izani (Universiti Kebangsaan Malaysia)

Local Arrangement and Logistic
Ts. Suziana Omar (UiTM)
Mohd Amir Hamzah Ab. Ghani (UiTM)
Rina Abdullah (UiTM)
Nik Nur Shaadah Nik Dzulkefli (UiTM)
Raja Mohd Noorhafizi Raja Daud (UiTM)

Master of Ceramony
Ir. Dr. Malathy Batumalay (INTI International University)

PM23 Conference

Journal of Physics: Conference Series **2627** (2023) 011002 doi:10.1088/1742-6596/2627/1/011002

Peer Review Statement

All papers published in this volume have been reviewed through processes administered by the Editors. Reviews were conducted by expert referees to the professional and scientific standards expected of a proceedings journal published by IOP Publishing.

- **Type of peer review:** Single Anonymous
- **Conference submission management system:** Morressier
- **Number of submissions received:** 30
- **Number of submissions sent for review:** 18
- **Number of submissions accepted:** 18
- **Acceptance Rate (Submissions Accepted / Submissions Received × 100):** 60
- **Average number of reviews per paper:** 1
- **Total number of reviewers involved:** 20
- **Contact person for queries:**
  Name: Zulzilawati Jusoh
  Email: zulzilawati@uitm.edu.my
  Affiliation: Universiti Teknologi MARA (UiTM) Cawangan Terengganu

Content from this work may be used under the terms of the Creative Commons Attribution 3.0 licence. Any further distribution of this work must maintain attribution to the author(s) and the title of the work, journal citation and DOI.
Published under licence by IOP Publishing Ltd

Design of High Performance Triangular Fiber Bragg Grating Temperature Sensor using OptiSystem and OptiGrating Software

Nani Fadzlina Naim[1], Nur Ainie Tasnim Muhamad Zulkifli[1], Latifah Sarah Supian[2], Norsuzila Ya'acob[1], Azita Laily Yusof[1], Wan Norsyafizan W. Muhamad[1], Suzi Seroja Sarnin[1]

[1]School of Electrical Engineering, College of Engineering, Universiti Teknologi MARA (UiTM), 40450 Shah Alam, Selangor.
[2]Department of Electrical & Electronics Engineering, Faculty of Engineering, National Defence University of Malaysia, Kuala Lumpur, Malaysia.

*Corresponding author: nanifadzlina@uitm.edu.my

Abstract. This paper presents on the design of high performance triangular Fiber Bragg Grating (FBG) as temperature sensor. This triangular FBG is designed and simulated using OptiGrating Software. The Triangular FBG design is exported to OptiSystem Software. In the OptiSystem software, the overall triangular FBG temperature sensor system is designed and simulated. It found that the triangular FBG has higher temperature sensitivity which is 37.5pm/°C than the standard FBG which is around 13pm/°C. It also has high linearity of 99.2% which specifies the consistency of wavelength shift as the temperature is increased. It also has higher resolution due to its sharp peak and smaller bandwidth.

1. Introduction

Optical fiber sensors based on fiber Bragg gratings (FBG) have attracted a lot of attention in recent years, and they are now becoming an industry standard for a wide range of applications in important industries including transportation, energy, safety, security, and medicine. In terms of tolerance to electromagnetic interference, dynamic response, multiplexing, and embedding capabilities, they implement multiple sensors, making them ideal for smart structures and structural health monitoring [1-2]. Fiber Bragg gratings (FBGs) are extensively utilized in a broad range of applications, spanning from optical communications to optical sensing, and have a long history [3]. FBG based sensors have shown to be an effective option for monitoring tasks in hazardous environments. Because of its self-referencing capabilities, large-scale multiplexing capabilities, and tolerance to electromagnetic interference, FBG is being recognized as an essential sensor technology. Wavelength division multiplexing (WDM) and time-division multiplexing (TDM) techniques have been widely applied to FBG arrays. However, they have restrictions in terms of optical bandwidth and low signal-to-noise ratio (SNR) which will limit the resolution and multiplexing abilities. WDM is a technique using a sequence of high reflectivity narrow bandwidth FBGs with different center wavelengths and suitable frequency spacing. A broadband laser source is used to probe these FBGs [4].

Content from this work may be used under the terms of the Creative Commons Attribution 3.0 licence. Any further distribution of this work must maintain attribution to the author(s) and the title of the work, journal citation and DOI.
Published under licence by IOP Publishing Ltd

In contrast, TDM uses a single narrowband laser source to interrogate a sequence of FBGs with identical center wavelengths. The TDM FBGs typically have low reflectivity along the optical cable. Using an optical to electrical converter, the reflected signals from the FBGs are transformed to electrical signals [5]. WDM is a popular technique due to its intuitionistic wavelength demodulation method. With a limited bandwidth of less than 100 nm for broadband light sources, typically only tens of sensors can be effectively multiplexed in one fiber. In TDM, equal wavelength FBG sensors with different time delay between reflected pulses along the fiber are detected. Thus, it is more potential for higher number of multiplexing sensors [5]. In sensor applications, the FBG is employed in the sensor head to detect changes such as strain and temperature [6].

The combination of OptiSystem, Optiwave's integrated solutions, and the inclusion of OptiGrating expands the application potential of FBGs and enhances their use in fiber optic sensing applications [7]. Using OptiGrating, FBG can be designed and characterized. The integration of OptiGrating and OptiSystem allows a wider application scenario [8]. The design and construction of FBG have lately gotten a lot of interest in the field of fiber optics. Recently, it has been proved that a new class of continuous evolutionary algorithms, known as covariance matrix adapted evolution strategy (CMAES), is successful in handling strongly linked multimodal optimization issues. In the optimization process, CMAES creates new population members by sampling from a probability distribution that was computed earlier in the process. Researchers in [9] studied two different design issues in order to illustrate the efficiencyof the CMAES algorithm in the design of triangular FBG filters. The design of an FBG filter with a particular bandwidth is taken into consideration.

In this project, triangular FBG is designed using OptiGrating software and the design is exported to OptiSystem software to validate the design. In OptiSystem software, the overall setup for FBG temperature sensor is designed and simulated. By deploying triangular FBG temperature sensor, it is found that the triangular FBG sensor has better performance in terms of resolution and sensitivity.

2. Research Methodology

In this paper, the design of triangular FBG is simulated using OptiGrating software. The OptiGrating software was used to construct and construct triangular FBG sensors under a variety of temperature conditions. The developed FBG is exported as spectrum files that were readable by the OptiSystem tool, allowing them to be used in the sensor setup that was designed using OptiSystem software. The OptiGrating technique is utilized to create triangular FBG sensors with a bandwidth of 0.2 nm, a total chirp of 2 nm, a length of 50mm, and an index modulation of $9.1 \times 10^{-0.05}$. After conducting the simulation using OptiGrating software, the design in exported to OptiSystem software. By varying the temperature, the FBG reflection spectrum is observed both using the OptiGrating and Optisystem software.

Figure 1 shows the FBG grating parameters that has been used for this design. Figure 2 shows the reflection and transmission spectrum of the FBG that has been simulated using Optigrating software. The blue line shows the triangular FBG reflection spectrum while the red line shows the transmission spectrum of the FBG. Figure 3 shows the grating index change versus the length of the FBG. This design is exported to OptiSystem software to be tested as FBG temperature sensor.

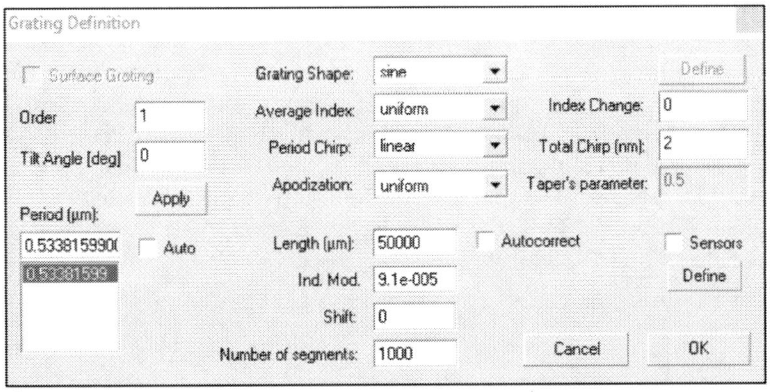

Figure 1. Grating parameters used in OptiGrating software

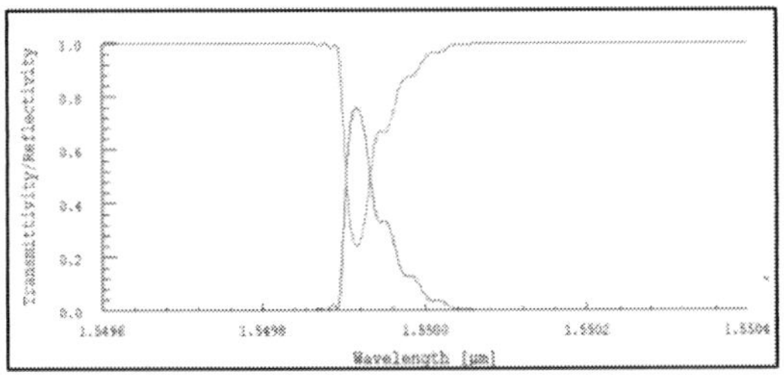

Figure 2. Reflectivity and transmittivity

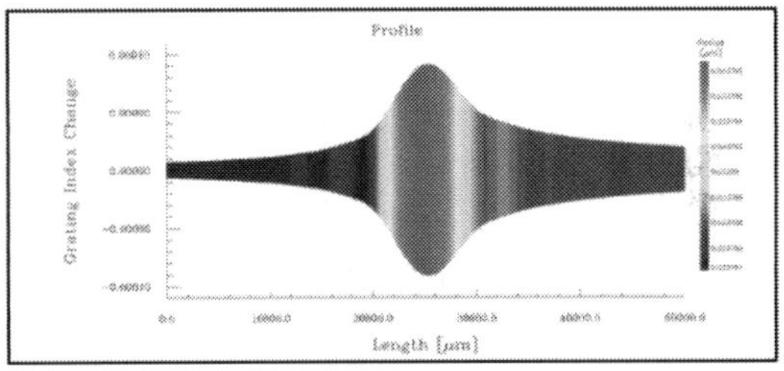

Figure 3. Grating Index Change versus length of the triangular FBG centered at 1549nm

A comparison of the linearity response of both of the developed FBG sensors as well as their sensitivity to the applied physical measurand was made. When it comes to temperature sensors, the first FBG, which is centered at 1549 nm, is devoted to non-temperature sensor measurements, and will be referred to as FBG_1, while the second FBG, which is centered at 1550 nm and is exposed to temperature fluctuations, will be referred to as FBG_2. Figure 2 shows the Grating Index change of the triangular FBG. The FBG temperature simulation model is represented in Figure 4 which includes white light source with wavelengths ranging from 1549 nm to 1553 nm, an optical circulator, single-mode optical fibers with lengths of 1 km and two triangular FBG sensors. The optical sources, each with a power of 5 dBm on average, are launched into the optical fiber through the circulator, where the reflected signal from the two FBG sensors, both designed by OptiGrating, is directed to the optical spectrum analyzers for analysis and decoding of the response signal. The two FBG_1 and FBG_2 that have been created by OptiGrating, as previously indicated in the preceding part, are uploaded into OptiSystem for further processing. In order to distinguish between the cross-sensitivity of the measurand, the FBG_1 is assumed to have no temperature sensor applied, whereas the FBG_2 is considered to have a temperature sensor applied.

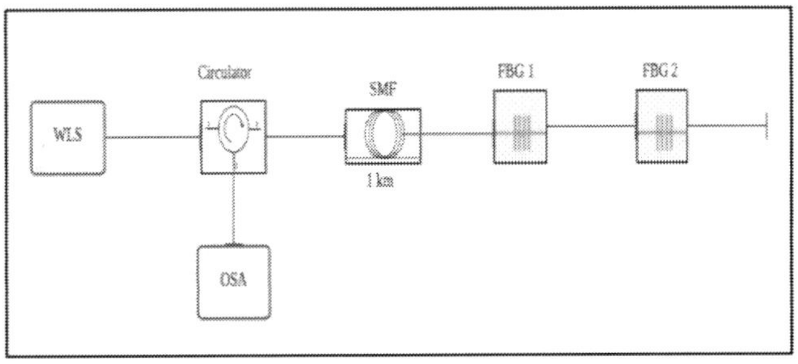

Figure 4. FBG Interrogation scheme as temperature sensor whereas WLS is White Light Source, SMF is Single Mode Fiber, OSA is Optical Spectrum Analyzer and FBG is Fiber Bragg Grating.

3. Result and Discussion

The simulation results are obtained using OptiGrating and OptiSystem softwares. The temperature is varying in the OptiGrating software and the FBG reflection spectrum is observed in OptiGrating and OptiSystem software. Figure 5 shows the wavelength shift versus the applied temperature. The blue line graph shows the simulation result of the wavelength shift from the triangular FBG when various temperature is applied. From both the theoretical and simulation results, it is found that the wavelength increases as the temperature increase. The simulation result shows higher sensitivity of 37.5pm/°C than the standard FBG which is around 13pm/°C. It also has high linearity of 99.2% which specifies the consistency of wavelength shift as the temperature is increased.

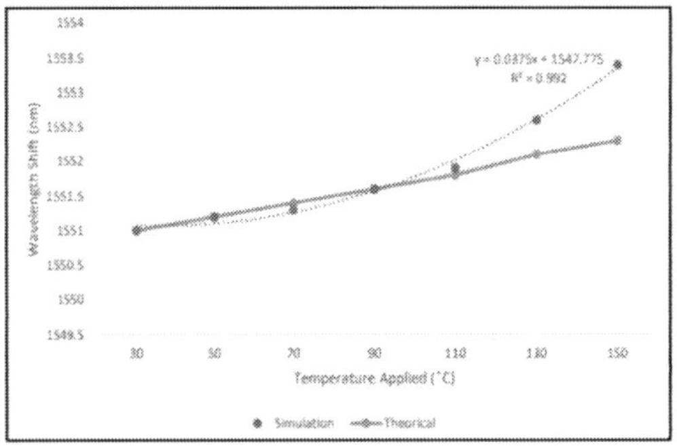

Figure 5. Comparison between theory and simulation of FBG wavelength shift for various temperature

Figure 5 shows the reflection spectrum of normal FBG temperature sensor as the temperature increase from 30°C to 150°C. The uniform FBG sensor has wider bandwidth as compared to the triangular FBG which will limit the resolution of the temperature sensor. Figure 6 shows the simulation result of the reflection spectrum of the triangular FBG as the temperature increase. Triangular FBG has sharp peak which will produce higher resolution of temperature sensor as compared to uniform or normal FBG.

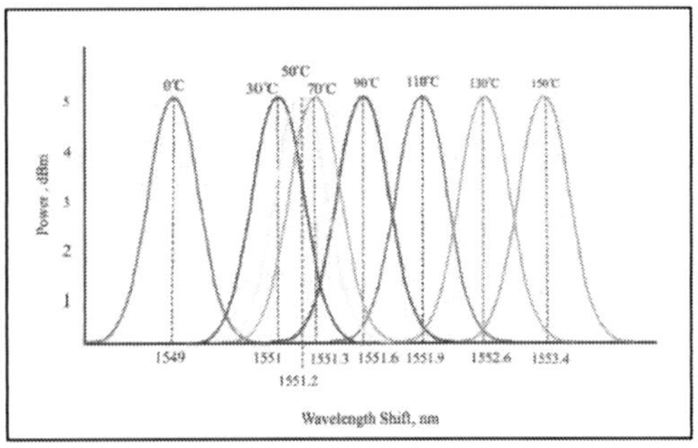

Figure 6. Reflection spectrum of uniform FBG

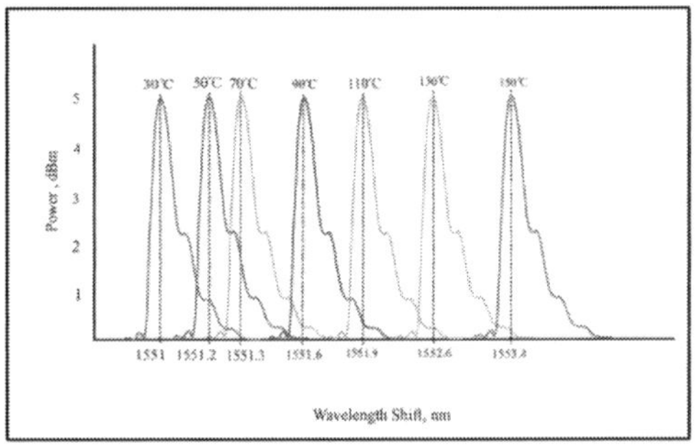

Figure 7. Reflection spectrum of triangular FBG

4. Conclusion

In conclusion, triangular FBG as temperature sensor has been successfully designed and simulated. Triangular FBG as temperature sensor has been designed using OptiGrating and the design is exported to OptiSystem software. In the OptiSystem software, the FBG temperature setup is designed and simulated. It is found that the triangular FBG has higher resolution and higher sensitivity of 37.5pm/°C than the standard FBG which is around 13pm/°C. It also has high linearity of 99.2% which specifies the consistency of wavelength shift as the temperature is increased.

References

[1] Zaidi, F., Nannipieri, T., Signorini, A., Taki, M., Donzella, V., & Di Pasquale, F. (2013). High performance time domain FBG dynamic interrogation scheme based on pulse coding. IEEE Photonics Technology Letters, 25(5), 460–463. https://doi.org/10.1109/LPT.2013.2241046

[2] Shadab, A., Jurel, N., & Sarswat, P. (2018). *Design & Analysis the parameters of strain based FBG sensors using Optigrating*. 4803–4811.

[3] Baskar, S., Suganthan, P. N., Ngo, N. Q., Alphones, A., & Zheng, R. T. (2006). Design of triangular FBG filter for sensor applications using covariance matrix adapted evolution algorithm. *Optics Communications*, *260*(2), 716–722. https://doi.org/10.1016/j.optcom.2005.11.037

[4] Hassan, K. N. A. K., Elgaud, M., Su'Ait, M. S., Bakar, A. A. A., & Zan, M. S. D. (2018). Signal to Noise Improvement Ratio of TDM-FBG Sensor Based on Golay Complementary Codes. 2018 IEEE 7th International Conference on Photonics, ICP 2018, 6–8. https://doi.org/10.1109/ICP.2018.8533179

[5] Wang, Z., Wen, H., Luo, Z., & Dai, Y. (2016). Time division multiplexing of 106 weak fiber Bragg gratings using a ring cavity configuration. Photonic Sensors, 6(2), 132–136. https://doi.org/10.1007/s13320-016-0275-2

[6] Huang, R., Zhou, Y., Cai, H., Qu, R., & Fang, Z. (2004). A fiber Bragg grating with triangular spectrum as wavelength readout in sensor systems. *Optics Communications*, *229*(1–6), 197–201. https://doi.org/10.1016/j.optcom.2003.10.045

[7] Naim, N. F., Ab-Rahman, M. S., Bakarman, H. A., & Bakar, A. A. A. (2013). Real-time monitoring in passive optical networks using a superluminescent LED with uniform and phase-shifted fiber Bragg gratings. *Journal of Optical Communications and Networking*, *5*(12), 1425–1430. https://doi.org/10.1364/JOCN.5.001425

[8] Elgaud, M. M., Zan, M. S. D., Abushagur, A. A. G., & Bakar, A. A. A. (2016). Analysis of independent strain-temperature fiber Bragg grating sensing technique using OptiSystem and OptiGrating. *2016 IEEE 6th International Conference on Photonics, ICP 2016, September 2018.*

[9] S. Baskar; P.N. Suganthan; N.Q. Ngo; A. Alphones; R.T. Zheng (2006). Design of triangular FBG filter for sensor applications using covariance matrix adapted evolution algorithm. *Optics Communications*, 260(2), 716–722.

Acknowledgments

The author wishes to thank Universiti Teknologi MARA (UiTM) under internal research grant 600-RMC/GPK 5/3 (101/2020) for supporting this project.

ASE from polymer films embedded with silver nanowires

Tamim Abdullah Altamimi[1], **Wan Maryam Wan Ahmad Kamil**[1*]

[1]School of Physics, Universiti Sains Malaysia, 11800 USM, Penang, Malaysia

*Corresponding author: wanmaryam@usm.my

Abstract Silver nanowires (AgNWs) were developed on glass substrate using polyol method at 160 °C. Centrifuge at 1500 rpm for different ratio of ethanol was done to extract the AgNWs. Rhodamine 6G (R6G) was then applied at different thicknesses. Enhanced emission from R6G was observed when the concentrations used are at 40 μL and 20 μL. Overall results indicate promising potential for enhancing plasmon resonances in optical emission by utilising silver nanowires as scatterers and R6G as the gain medium.

1. Introduction

A random laser is a type of laser that occurs when photons remain in a semiconductor long enough to achieve gain without the use of mirrors. The photons are contained by a set of mirrors in a conventional laser configuration. However, a random laser has no reflectors. They confine photons by maintaining internal scattering from nanowire structures long enough within the gain medium for photons to achieve gain for lasing. This is possible by altering the refractive index within the gain medium; in this case using silver nanowires (AgNWs). Due to the optical cavity provided by nanowires, lasing is achievable despite the limited quantity of gain material. Since its first presentation in 2001, significant scientific effort has been put into the development of semiconductor nanostructures, which are sometimes referred to as nanowire lasers [1]. Polymer films containing nanoparticles of silver and weakly scattering polymers have produced random lasers having coherent feedback [2,3]. A random laser was made by placing silver nanoflowers on a polymer membrane and placing the membrane on a flexible substrate [4]. Using a fluoresecent laser dye (especially DCM)-doped polymer sheet with Zinc oxide (ZnO) micro-cabbages as scatterers, a system with low-threshold continuous-wave random laser is reported [5]. It was reported that the threshold for random lasing may be greatly lowered by using a plasmonic hybrid structure. An example is using gold–silica (Au core-SiO_2) shell nanocubes coated on silver (Ag) film (Au@SiO_2 NCs-Ag film) [6]. However, all these works rely on liquid-based samples. In this paper, solid-state gain medium of Rhodamine 6G with the AgNWs is attempted towards making a solid-based random laser device suitable for integration with current laser systems.

In this work, AgNWs was synthesized with a polyol method. The structural and optical properties of AgNWs coated with R6G as the gain medium are discussed. The AgNWs developed using polyols as a simple, rapid, and environmentally friendly approach is proposed. This method produces AgNWs with substantial aspect ratio. For the purpose of finding the best gain medium thickness, reaction parameters on the synthesis process, such as temperature range, molecular of the polyvinylpyrrolidone (PVP) molecular weight, Sodium chloride (NaCl)

Content from this work may be used under the terms of the Creative Commons Attribution 3.0 licence. Any further distribution of this work must maintain attribution to the author(s) and the title of the work, journal citation and DOI.
Published under licence by IOP Publishing Ltd

concentration, and mixing rate of the solution, were kept constant in the preparation of the silver nanowires. The goal of this study was to investigate the effect of the dye amount thickness prepared with same concentration of dye drop-casted on to AgNWs on the emission of R6G.

2. Preparation Method

The Ag NWs was synthesized using a modified polyol procedure. Solutions of 0.45 molarity Polyvinylpyrrolidone (PVP), 0.2 molarity Sodium chloride (NaCl), 0.2 molarity Potassium bromide (KBr), as well as 0.095 molarity of silver nitrate ($AgNO_3$) were prepared. In a 50 mL flask, 12 mL of ethylene glycol (EG) was added and warmed in an oil bath for 10 minutes prior to the addition of 0.02 g of NaCl, and 0.04 g of KBr. Then with magnetic stirring at 250 rpm, 0.35 g of the PVP was added. After 10 minutes of gentle stirring, 0.14 g of the silver nitrate was added into the solution. After the completion of the added, the reaction was kept till the color was changed to a bluish green. Deionized water (DI) was used to dilute the samples at a ratio of 1:1 and left to cool at room temperature. The speed of centrifuge was at 1500 rpm to purify the AgNWs whereby the heavy particles were precipitate, and the light nanowires would be on the top of the liquid (supernatant). The supernatant was washed 3 times with acetone as well as ethanol before being reconstituted in 10 ml of ethanol for further analysis.

The Ag nanowire suspended in ethanol is coated on glass substrate and dried on a hot plate at approximately 60 °C. Next, R6G prepared in ethanol with concentration of 2.088mM was mixed with (4.4 mg/ml) of polyvinylpyrrolidone polymer (PVP) and coated on the glass substrates above the silver nanowires at different thicknesses corresponds to each concentration of 20μL to 120μL and left to dry at room temperature for 24 hours.

For information on morphology and structural characteristics on the sample, field emission scanning electron microscope (FESEM) measurements were conducted. The Optical properties of R6G on AgNWs were investigated by photoluminescence measurements. A 355 nm Nd:YAG laser source with 0.6 ns pulse width and 20 kHz repetition rate was employed to excite the samples. The emission was recorded using Ocean optics fibre spectrometer and displayed the spectrum was using standard Ocean View software on standard PC.

3. Results and discussion

3.1 Morphology and structural characteristics

Fig. 1. shows FESEM images of the surface form for all samples with different amount of R6G drop casted on the AgNWs. The average diameter and length of AgNWs were analysed using ImageJ software. The average length and diameter of AgNWs was about 4.6 μm, and 16 nm respectively, before and after inclusion of the dye. No changes of the length and diameter was observed. However, the amount of AgNWs observed on the surface decreases as the thickness of the dye increased.

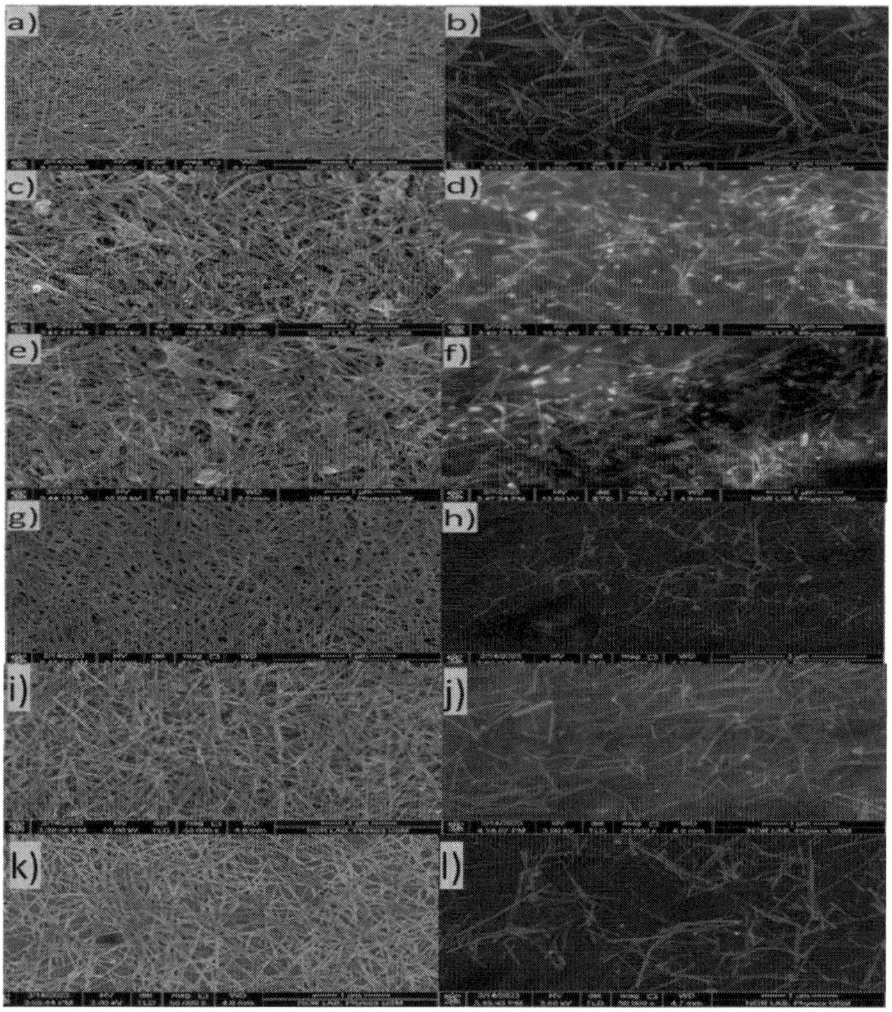

Fig.1. (a) AgNWs before Doping R6G. (b)AgNWs after Dye, thickness corresponds to concentration 20μL, (c)AgNWs before Doping R6G. (d) AgNWs after Dye, thickness corresponds to concentration 40μL, (e) AgNWs before Doping R6G. (f) AgNWs after Dye, thickness corresponds to concentration 60μL. (g) AgNWs before Doping R6G. (h) AgNWs after Dye, thickness corresponds to concentration 80μL, (i) AgNWs before Doping R6G. (j) AgNWs after Dye, thickness corresponds to concentration 100μL and, (k) AgNWs before Doping R6G. (l) AgNWs after Dye, thickness corresponds to concentration 120μL.

3.2 R6G emission

Photoluminescence measurements were performed on all samples at a constant pump power of 31 mW and is displayed in Fig.2. The highest peak marked as yellow represent the lowest thickness of R6G drop cast onto the sample. As the thickness amount increases, the peaks decrease. However, by measuring the peak's full width at half maximum (FWHM), the green peak showed the lowest FWHM of 57.7 nm as compared to 74.2 nm obtained from the yellow peak. This indicates spectral narrowing of the green peak when the R6G was 40µL, which indicates an onset of amplified spontaneous emission (ASE) [7].

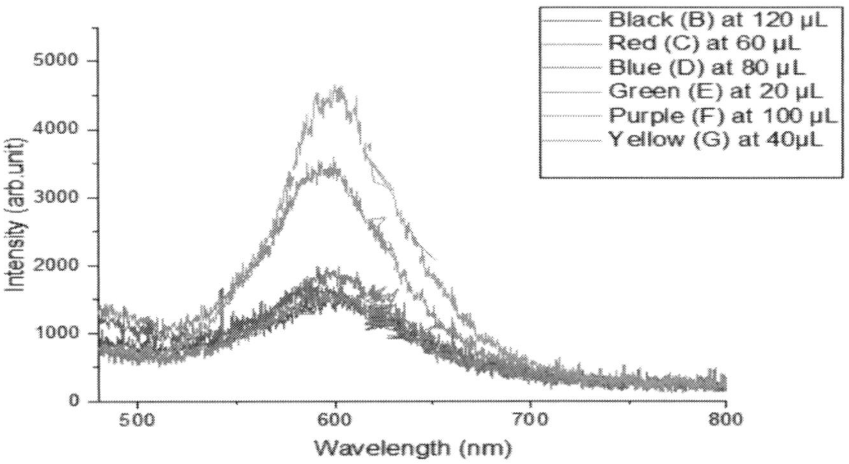

Fig.2. Intensity Vs Wavelength samples of AgNWs with different dye thicknesses.

Table.1, summarizes key parameters such as the FWHM, peak centre wavelength and peak height for all samples starting from highest peak to lowest peak indicated by Yellow (G) - Green(E)- Red(C)- Blue(D) – Purple(F) – Black(B) curves in Fig. 2. The yellow peak represent the lowest thickness corresponds to amount of 20µL of dye and black is the thickest dye corresponds to amount of 120 µL. As shown in table. 1, the centre peak wavelength is slightly shifted in two thicknesses of dye whereby the yellow peak is shifted to higher wavelength and the purple is shifted to lower wavelength. This is consistent with reported work due to the different amounts of dye [8]. By comparing the FESEM images in Fig.1, the thickness of R6G influenced the amount of AgNWs appearing on the surface. The amount of NWs present at 40 µL of R6G is adequate to create ASE. To improve the emission, more work is required to refine the concentration and thickness of R6G so as to allow more AgNWs surface but at the same time provide sufficient gain from R6G to observe ASE and also lasing.

Table 1. R6G with different thicknesses to measure FWHM, peak center and, peak height.

R6G (µL)	FWHM (nm)	Peak center (nm)	Peak height (arb.unit)
20	74.24	599.94	4555.82
40	57.72	591.44	3877.87
60	80.18	591.44	1896
80	84.76	591.44	1821.42
100	76.71	585.64	1656.45
120	97.30	591.44	1590.92

4. Conclusion

Using AgNWs as metal particles for enhancing luminescence emission by means of plasmonic resonances with various dye thicknesses demonstrates ASE at 40 µL of R6G. Shifts in wavelength occurs when thickness is below of concentration 40 µL and above 100 µL. Low thickness of dyes showed enhanced luminescence due to the amount of AgNWs present at the surface of the sample.

References

1. M.H. Huang, S. Mao, H. Feick, H. Yan, Y. Wu, H. Kind, E. Weber, R. Russo and P. Yang, Room-temperature ultraviolet nanowire lasers, Nature 292, 1897 (2001).

2. Meng, X., Fujita, K., Zong, Y., Murai, S., & Tanaka, K. (2008). Random lasers with coherent feedback from highly transparent polymer films embedded with silver nanoparticles. Applied Physics Letters, 92(20), 201112.

3. Meng, X., Fujita, K., Murai, S., & Tanaka, K. (2009). Coherent random lasers in weakly scattering polymer films containing silver nanoparticles. Physical Review A, 79(5).

4. Tong, J., Li, S., Chen, C., Fu, Y., Cao, F., Niu, L., . . . Zhang, X. (2019). Flexible Random Laser Using Silver Nanoflowers. Polymers, 11(4), 619.

5. Pramanik, A., Biswas, S., Kumbhakar, P., & Kumbhakar, P. (2021). External feedback assisted reduction of the lasing threshold of a continuous wave random laser in a dye doped polymer film and demonstration of speckle free imaging. Journal of Luminescence, 230, 117720.

6. Ning, S., Dai, K., Zhang, N., Zhang, Y., Wu, Y., Huang, J., . . . Zhang, F. (2021). Improving the random lasing performance using Au@SiO2 nanocubes-silver film hybrid structure. Journal of Luminescence, 231, 117788.

7. Christianson, N. D., Lu, Y., & Dawson, N. J. (2019, December). Recovery of photodegraded rhodamine 6g in ester-containing polymer matrices. *Photochemical & Photobiological Sciences*, *18*(12), 2865–2874.

8. Maryam, W., Tan, H. H., Jagadish, C., Dawes, J. M., Zhao, B., & Wan Ismail, W. (2021, December 22). A hybrid random laser using dye with self-organized GaN nanorods. *Semiconductor Science and Technology*, *37*(2), 025009.

Design and modelling of modulated cw laser generation in lidar application

D A Zainal[1], **M M I Ahmeed**[1] **and L S Supian**[1]

[1] Department of Electrical and Electronics Engineering, Faculty of Engineering, National Defence University of Malaysia, Kem Sg Besi, 57000, Kuala Lumpur, Malaysia.

3201279@alfateh.upnm.edu.my

Abstract. In this work, we designed a configuration that may be used for direct detection of light, detection and ranging sensing method based on generated continuous wave laser and characterized by range measurements. The generation of modulated laser distance is studied and analysed by several distances where modulation of laser signal is utilized. The modelling configuration consisted of transmission module that comprises of signal generator, modulator, and propagation module, object target, and receiver modules comprises of photodetector, amplifier, filter, and visualizers. The design and configuration were developed by Optisystem layout, and the study and analysis are done by simulation using the developed model. A subsystem of range measurement was designed to detect varied distances of detection of target object by the generated modulation of continuous wave laser at 10, 50, 100 and 200 meters respectively. The results discussed were based on the power received and frequency given by different detection ranges. Analysis shows that the detection scheme is sensitive to the received electromagnetic radiation and the signal is better at shorter haul based on the generated modulation of the continuous wave source and the signal deteriorated as target distance is increased.

1. Introduction

Distance detection and measurement through free space using light detection and ranging functions by transmitting and receiving electromagnetic energy. Light detection and ranging or lidar, can be presented by pulse radar or continuous wave radar, where pulse radar may represent a simpler system, and continuous wave has a very high time bandwidth when it is frequency modulated [1]. In lidar, the signal generator or in transmission subsystem, the amplifier and laser determine the wavelength and waveform as well as the power of the transmitted signal. The receiver function will transform the propagating energy into an electrical signal that will be processed to extract the desired information [2]. The range and time are related by equation (1), where c is the speed of light:

$$T = 2d/c \tag{1}$$

In lidar, the system usually has smaller bandwidths than the target, whilst the target may be similar to a Lambertian or specular reflector or combination of both. There are many factors that may affect the reflected signal from the object target and back to the receiver module such as the laser module, environment factor, target reflectivity and angular dispersion [3]. Distance measurement by using modulated wave method have been utilized in many applications such as noncontact surface

Content from this work may be used under the terms of the Creative Commons Attribution 3.0 licence. Any further distribution of this work must maintain attribution to the author(s) and the title of the work, journal citation and DOI.
Published under licence by IOP Publishing Ltd

profiling [4], optical sensing [5] and positioning due to the advantages of optimum resolution in short distance sensing. The signal may generate saw-tooth bias current in the modulator of the wavelength-tunable laser diode. The performance of the lidar may be measured by signal to noise ratio by considering the noise sources that may exist in the system such as thermal noise, dark noise and shot noise. This work presents modeling configuration by simulation using Optisystem for lidar by modulated continuous laser by analyzing the waveforms, detection signal and power by varied distances.

2. Design configurations

A configuration of lidar detection system was designed using modulated continuous wave by Optisystem simulation. The main subsystems are transmission module that are made up of signal generator, modulator, laser source, amplifier, where the signal is processed, and the propagating signal hit the target object and the receiver module that consists of photodetector, amplifier and filter processed the received signal. The built configuration is shown in Figure 1.

The carrier signal is produced by signal generator with frequency bandwidth of 300 MHz, and it is modulated with continuous wave laser. Modulator in transmission module modulates the signal generated by frequency modulation with continuous wave source. The modulated signal is then amplified, and the optical gain amplifier emitted the signal through where the signal then hit the target object. Parameters are set in the simulation for reflection process in terms of time. Varied distances can be set for each test and different frequencies may be observed after optical-electrical conversion.

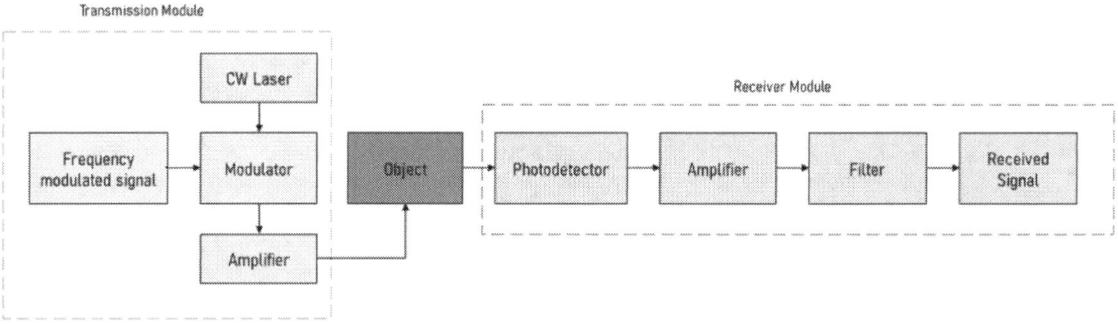

Figure 1. Lidar configuration of modulated continuous wave.

Several distances were set in the parameter layout from the transmission module to the target object, which in this work, the range of distances are set as d_n = 10, 50, 100, and 200 meters. Optical delayer that is represented by time delay in the configuration is used and to obtain different distance from transmission module to the target object, the delayer parameters are varied. The reflected signal from the target is received by photodetector and amplified before being filtered and processed. After the signal passed through the filter, signal received is observed in frequency domain and time domain.

This work observed the differences of frequencies and signal power of the received signal by varying the distances of the target object and observed the sensitivity of the system in term of varied distance range. The reflected signal power can be determined analytically by equation (2), where the measurement considers the parameters such as transmission loss, atmospheric loss factor, object or target reflection and area and distance of the object or target.

$$Power_{rec} = \frac{P_{trans}T_{ref}t_{opt}A^2t_{atm}^2}{4dn^2a_{ill}} \quad (2)$$

3. Results and discussion

By referring to the configuration as shown in Figure 1 above, we simulated the model using modulated continuous laser where the wavelength set is around 1552 nm, frequency of 300 MHz, power of 40 mW, threshold current of 40 mA, and amplifier of gain 20 dB. After setting the parameters as stated, we run the simulation at set distance ranges, d_n = 10, 50, 100, 200 meters. The results obtained are shown for d_n of 10, 50, 100 and 200 meters respectively from the target.

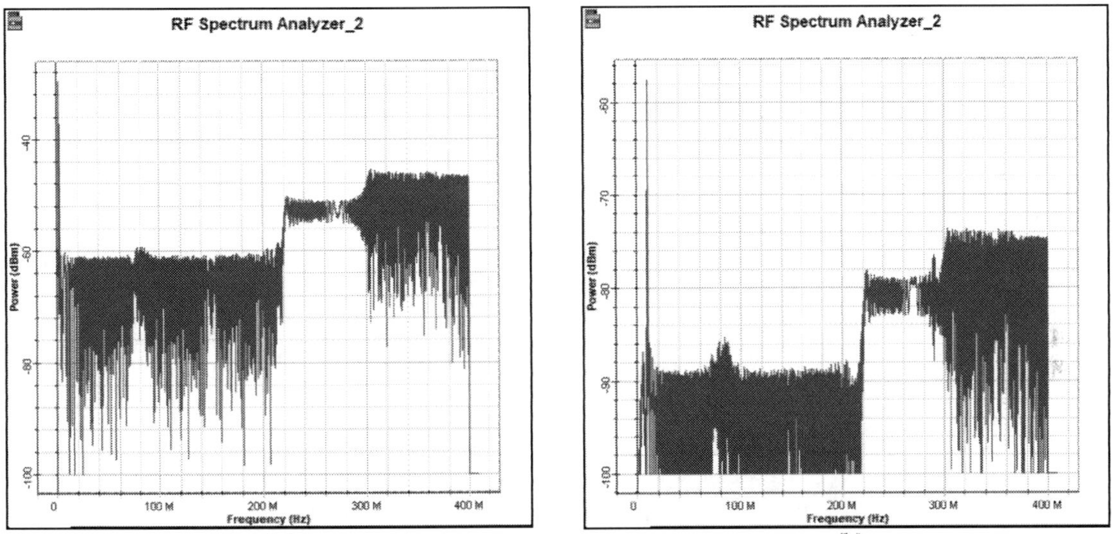

(a) (b)

Figure 2. The frequency spectrum at d_n of (a) 10 meters and (b) 50 meters respectively.

As shown in Figure 2, the distances set for the simulation model is 10 meters and 50 meters from the object target respectively where the frequency spectrum waveforms are depicted. The signal power is slightly lower for CW signal of distance of 10 meters compared 50 meters. The frequency spectrum of d_n at 10 meters is around 8.00×10^6 Hz and 1.02×10^7 Hz for d_n of 50 meters.

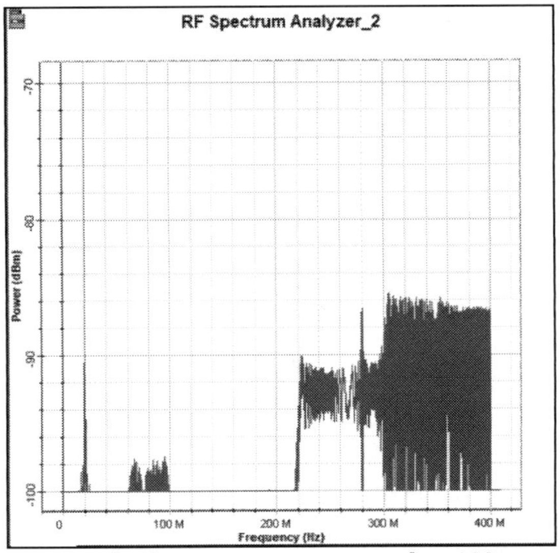

Figure 3. The frequency spectrum at d_n at 100 meters.

On the other hands, when the distance is increased to 100 meters as shown in Figure 3, the signal power is increased whilst the frequency spectrum is also increased. Figure 4 shows the single peak of waveforms of target distance of 100 meters where the frequency spectrum is around 2.00×10^7 Hz.

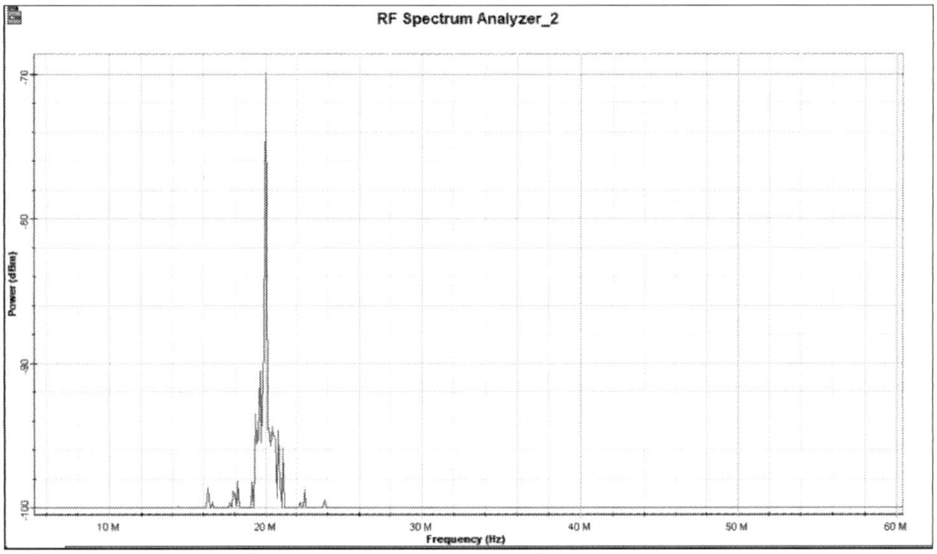

Figure 4. The frequency spectrum at d_n at 100 meters.

When the target distance is further increased to 200 meters, the signal power also increases as depicted in Figure 5. However, the target object is undetected when the distance is further increased to more than 450 meters. In the simulations done, thermal noise and shot noise are considered in the parameters.

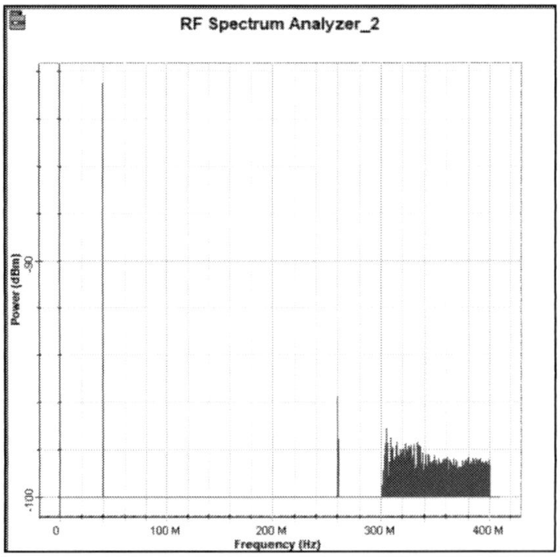

Figure 5. The frequency spectrum at d_n at 200 meters.

Journal of Physics: Conference Series **2627** (2023) 012003 doi:10.1088/1742-6596/2627/1/012003

Since this work was carried out using simulations only, the ideal conditions are expected. However, in real environment for lidar, many factors should be considered that may influence the target detection such as the atmospheric conditions, noise, and interferences from other signals or target.

4. Conclusions

In this work, we showed detection configuration and modelling for simple detection using continuous wave where the laser was frequency modulated and the system was characterized by the several target distances. We described the basic concept of frequency modulated light detection and ranging by target distance through simulation. From the results shown, continuous wave modulated laser shows good sensitivity and signal power received at the short-haul distance, however, the optimum range detection is quite limited to less than 500 meters. In real environment, the signal power and frequencies might deviate slightly due to interference and various conditions of the environment, atmospheric, and noises between the transmission and receiver components.

Acknowledgment

This work would not have been possible without the financial and facility support of the Fundamental Research Grant Scheme (FRGS/1/2017/STG07/UPNM/03/3), Ministry Higher Education of Malaysia and National Defence University of Malaysia.

References

[1] Qingchuan H, Kaida C, Jing X 2021 *J. of Physics: Conference Series.* **1939** 012066
[2] Richmond R D and Cain S C 2010 *Direct Detection LADAR Systems* (Washington-SPIE Press)
[3] Elghandour A H and Ren C D 2013 *Int. Symposium on Photoelectronic Detection and Imaging 2013: Laser Sensing and Imaging and Applications* **8905** pp1-9
[4] Amann M C, Bosch T, Myllyla R and Rioux M 2001 *Optical Engineering* **40(1)** pp 10-19
[5] Supian, L S, Zainal D A, Ahmeed M M I et al. 2023 *Int Conf of Future Sensing Tech* (Yokohama:SPIE)

Development and analysis of whispering gallery mode model for refractive index sensing

M. Batumalay[2] , H.A Zain [1], H.R.A. Rahim[3] and S. W. Harun[1]

[1] Department of Electrical Engineering, University of Malaya, 50603, Kuala Lumpur, Malaysia.

[2] Faculty of Data Science and Information Technology, INTI International University, 71800 Nilai, Negeri Sembilan, Malaysia.

[3] Faculty of Electronic and Computer Engineering, Universiti Teknikal Malaysia Melaka, Melaka, Malaysia.

malathy.batumalay@newinti.edu.my

Abstract. A refractive index sensor based on a whispering gallery mode resonator is numerically investigated in this paper. Whispering gallery mode resonators confine light and can be used in sensing applications, and this work presents the model built using COMSOL software to analyze their sensing performance. This work has investigated the WGM resonators' response to refractive index changes numerically using 2D models in the refractive index ranges 1.000 and 1.333. The resonator showed a sensitivity of 167 nm/RIU in the refractive index range of 1.3331.4 and a sensitivity of 95 nm/RIU in the refractive index range of 1.000-1.010. The sensing resonator has potential applications for biological and chemical analytes.

1. Introduction

The Micro-bottle resonators (MBR) are bottle-like structures fabricated by varying the radius of an optical fiber [1]. MBRs can support whispering gallery modes (WGM) due to the total internal reflection of the light coupled into the MBRs [2-4]. MBRs have a significant advantage in sensing and other advanced optical applications due to their light confinement abilities in a relatively small mode volume and high Q factor[5]. They can be used for various applications including environmental, chemical, or biological sensors [6]. For instance, hollow core silica capillary MBRs are used for magnetic field sensing and showed an 8.45 pm/Gs sensitivity with a Q factor of 1.28×10^6 [7]. A packaged optofluidic MBR for fluid flow rate sensing showed a maximum sensitivity of 0.079 pm/(μL/min) within 0–200 μL [8]. However, hollow MBRs sensors can become fragile and challenging.

Many works have been reported on WGM based sensors [9, 10]. The numerical analysis of WGM resonators showed a 3.14 mV/Pa sensitivity as an acoustic sensor. The broadband sensor was siliconbased [11]. A WGM gyroscope was shown to have a 3.002 pm/(°.s^{-1}) [12]. Dual measurements of refractive index changes and pressure were conducted using a WGM resonator. The sensor used two whispering gallery modes to achieve an RI sensitivity of 29.07 nm/RIU and a pressure sensitivity of 0.576 pm/kPa for the first mode, and an RI sensitivity of 38.68 nm/RIU and a pressure sensitivity of 0.589 pm/kPa for the other mode [13]. Dual temperature and refractive index measurements have also shown refractive index sensitivities of 45.8821 and 72.9402 nm/RIU and 0.0730 and 0.0703 nm/K temperature sensitivity [14]. Plasmonic whispering gallery modes have been shown to reach a 180

Content from this work may be used under the terms of the Creative Commons Attribution 3.0 licence. Any further distribution of this work must maintain attribution to the author(s) and the title of the work, journal citation and DOI.
Published under licence by IOP Publishing Ltd

nm/RIU sensitivity [15]. These numerical investigations show the potential whispering gallery mode resonators have as chemical, environmental, and biomedical sensors. Solid MBR sensors can have the analyte on the outer medium while a hollow MBR can have the analyte inside it or outside the MBR.

The sensing principle of MBR is to measure changes in the resonance wavelength shift or power output variations in the presence of the analyte. MBRs can be used in many sensing and measurement applications [2, 16, 17]. This model presents MBRs' azimuthal plane simulation for refractive index sensing. In MBRs, two projections exist, one in the (r, □) plane and the other in the (r, z) plane. The (r, □) plane for MBRs confines light like a whispering gallery. The light propagation and interactions with the changing surrounding medium are modelled and studied as a WGM resonator coupled with a straight waveguide. The effect of temperature on the transmission spectrum is also investigated to highlight the potential of MBR sensor to detect temperature change. Due to the complexity of the calculation, a 2D model is used to study the WGM resonator. Using a 3D model to simulate a WGM resonator coupled with a straight waveguide or fibre is computationally intensive and challenging to maintain due to their large size.

In this paper, the objective was to build a model in the (r, □) plane to study the MBR's sensing response. The model was used to investigate the resonator's response to refractive index changes in the ranges 1.333 and 1.000. The 2D model was also used to study the effects of temperature variations on the resonance wavelength shift.

2. Methodology and model development

The This section provides a detailed method for developing the COMSOL model for the WGM resonator. COMSOL 6.0 was used in this model. COMSOL solvers decompose the system into subdomains. The partial differential equations for each of the subdomains are solved with boundary conditions. The solutions are combined to finalize the system calculation. COMSOL can mesh and solve curved structures, making the software suitable for studying our WGM resonator. Other COMSOL wave optics modules simulate electromagnetic wave propagation and resonance phenomena. The module calculates electromagnetic field distributions and transmission coefficients for a photonics sensor design. The module's model formulas are based on Maxwell's equations combined with material properties for varied media propagation. The module solves Maxwell's equations in differential forms with initial/boundary conditions.

In this work, a model for a WGM resonator coupled with a waveguide was built using COMSOL 6.0. The simulation was conducted using electromagnetic waves and frequency domain analysis. A 2D structure was used to minimize the running time. The WGM resonator was tested with multiple radii from 3-11 μm. The waveguide was tested at radii 0.55-1.25 μm. The combination of WGM with a 3 μm radius and a 0.65 μm waveguide was chosen for further analysis. This combination was suitable for our study since it is optimum in terms of run time and mesh complexity. Larger resonator radii resulted in file sizes larger than 18 gigabytes for the model and computation times up to 12 hours for a single run of wavelength sweep with one value of the refractive index. The waveguide diameter was the value at which the transmission of the fiber has the minimum transmission at resonance dip.

The first step is using the model wizard. The dimension of the model is chosen as 2D. The physics chosen is the electromagnetics wave frequency domain (ewfd). The electric field of this model became (Ex, Ey, Ez). The WGM resonator radius and the waveguide diameter were used in the geometry definition. The refractive indices of the core, cladding, and analyte were used in the material definitions. A circle with a 7 μm radius was built in the geometry window. The circle represents the WGM resonator. A rectangle with a diameter of 0.65 μm was built. The rectangle represented the waveguide. The design has two elements: WGM resonator, and straight waveguide, and they were enclosed in a larger rectangle to form a model.

The model needed a perfectly matching layer to operate correctly. A perfectly matching layer mimics an open, non-reflecting, infinite domain. It is an alternative to non-reflecting boundary conditions. The layers work for all wave types. The perfectly matching layer was added to the boundary of the model to truncate the computations. The perfectly matching domain was used to avoid reflections. The layer absorbs the outgoing waves from the components inside the model without reflecting those waves into the elements of the model. The PML is applied to domains, not boundaries. Thus, domains need to be included in the model to use the PML. This model included the PML surrounding the encasement rectangle. The material of the PML was the same as the material of the encasement rectangle. Both materials had a 1.00 refractive index. The PML with cartesian coordinates as defined in the definition's tap in the components window. PML domains effectively weaken reflections. However, PML domains require high computational performance, as shown in Figure 1.

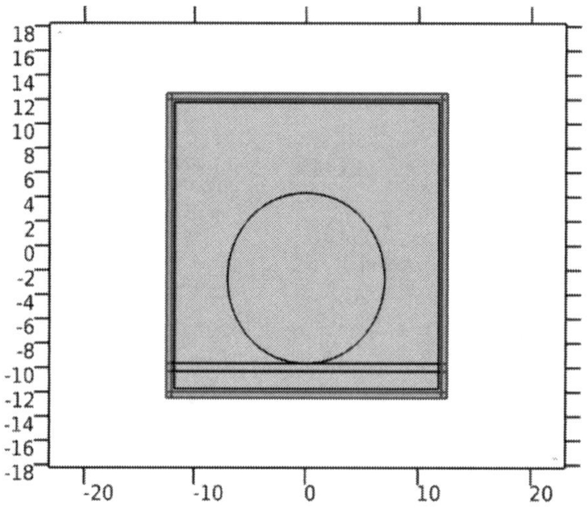

Figure 1. The PML configuration and geometry.

3. Results and discussion

This section presents the output performance of the proposed model. The input light was set within the 1550 nm range. The transmission spectra of the straight waveguide and the whispering gallery mode (WGM)-coupled straight waveguide (or tapered fiber) were investigated by sweeping the frequency. The effect of coupling gap distance on the performance of Q factor and extinction ratio was also calculated. The study was conducted within two refractive index ranges: 1.00-1.01 and 1.333-1.34. The resonance dip shifts were calculated and presented to quantify the potential of the sensor. The temperature response of the sensor was calculated to show the sensor's potential. The light was launched through port 1 and collected from port 2.

Figure 2 shows the transmission spectra of the straight waveguide (or fiber) and the WGM coupled fiber. It is observed that a resonance occurs in a WGM-coupled fiber. Far from the resonance range, most of the waveguide power remains in the waveguide. When the light propagating inside the fiber has a frequency and phase match with the resonator, the light gets coupled into the resonator instead of propagating through the fiber. Figure 3 shows the light field propagation inside the proposed WGMcoupled fiber at two resonance wavelengths of 1.545 and 1.564 μm. At the resonance range of 1.545 μm, the transmission spectrum show that the energy is coupled into the WGM resonator, see Figure 3.

A WGM of the second radial order occurs at this wavelength. At the resonance range of 1.545 µm. The power coupling between the WGM and the waveguide is apparent in both figures.

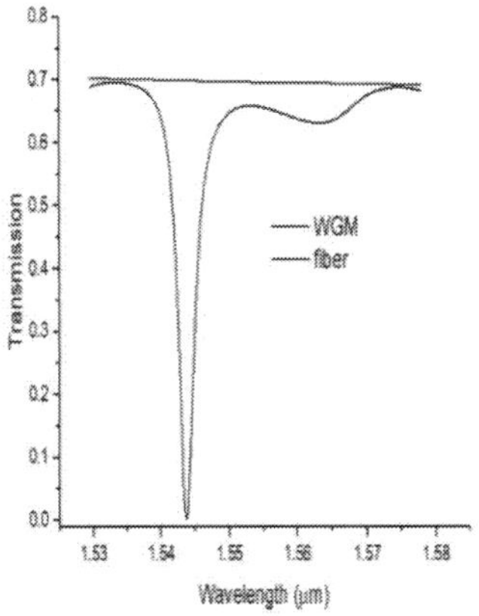

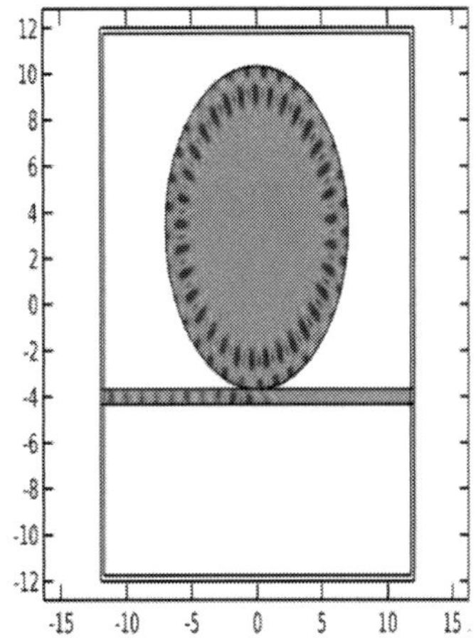

Figure 2. The transmission **Figure 3.** The resonance spectra of the fiber and
the response at 1.545 µm.
WGM-coupled fiber

The resonator can sense biochemical analytes by measuring the surrounding medium refractive index changes. These biological and chemical concentration changes induce refractive index changes in the resonator's surroundings. The fields of the resonator interact with the analytes, and a subsequent resonance wavelength shift occurs. This shifting is the change in the resonant wavelength of MBR. This wavelength shift can be tracked and used to detect and measure analytes. The resonant wavelength shift can be measured directly or indirectly. These resonance shifts can occur when the surrounding medium has a refractive index change or if the resonator's size or refractive index changes. When the MBR has a particle with a higher refractive index attached to its surface, part of the electric field is outwardly pulled. Thus, this particle causes the optical path to be longer, and a redshift happens to the resonance mode.

The sensitivity of this reaction is the ratio of the resonance wavelength shift to the refractive index change. The refractive index of the surrounding media directly affects the coupling conditions of the resonator and its transmission spectrum, as shown in Figure 5. With higher refractive indices, a red shift occurs in the WGM transmission spectra. The refractive index was changed within a range from 1.0001.010 with a step size of 0.0025. Figure 4 shows the resonance wavelength shifting towards longer wavelengths with increasing refractive index.

The model was then used to measure the sensitivity of the proposed WGM setup as the surrounding refractive index was changed in a range of 1.000-1.010 for gas sensing. The resonance wavelength experiences a redshift when the test gas is applied to the resonator surface. Figure 5 shows the sensing response, which indicates the sensitivity of 95 nm/RIU with a linearity of 100%.

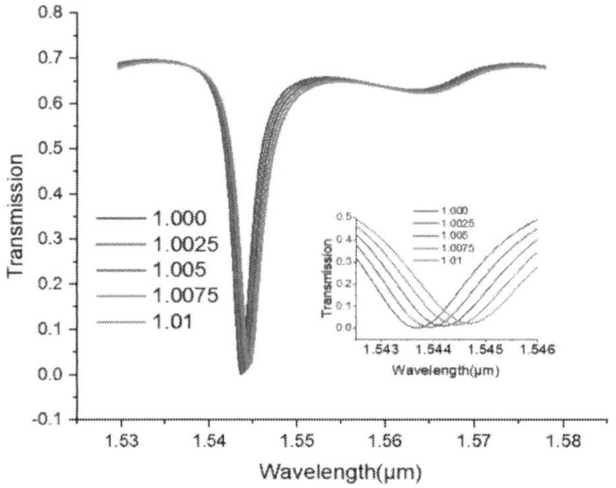

Figure 4. The transmission spectra at different refractive indices.

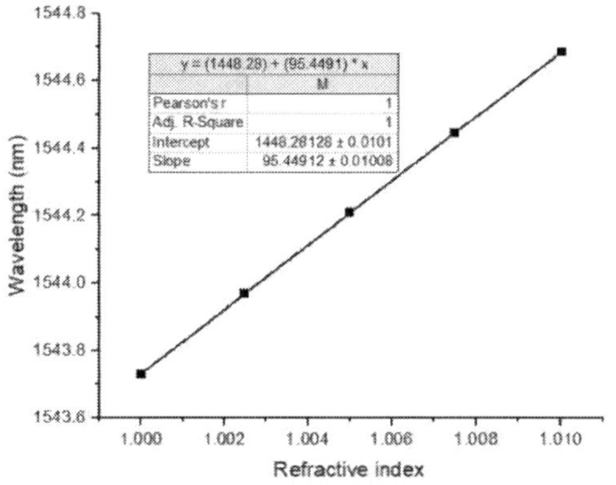

Figure 5. The sensitivity of the resonator at the refractive index range 1-1.01.

The shift in WGM resonance wavelength (or frequency) is not only caused by the change in the resonator size and refractive index of the resonator material, but also attributed to the change in the refractive index of surrounding medium [18]. Here, the refractive index range of 1.333-1.34 represents the range of various liquid solutions. Sodium nitrate, formalin, and ethanol solution have been shown to have refractive index values in this range. A test layer was added to the model to represent the liquid solution. Then, frequency sweep is applied to obtain the transmission spectra of the resonator. Figure 6 shows the transmission spectra at various liquid refractive indexes. As the surrounding media's refractive index is increased, the resonance dip experiences a redshift. The resonance dip moves to a longer wavelength due to the change in the coupling condition. It experienced a shift from 1.629 to 1.641 μm with the increase of refractive index from 1.333 to 1.400. Figure 7 shows a plot of resonance wavelength as a function of the surrounding refractive index. It shows a linear response with a sensitivity of 167.8 nm/RIU. The modelling results confirm the WGM resonance shift induced by the surrounding medium.

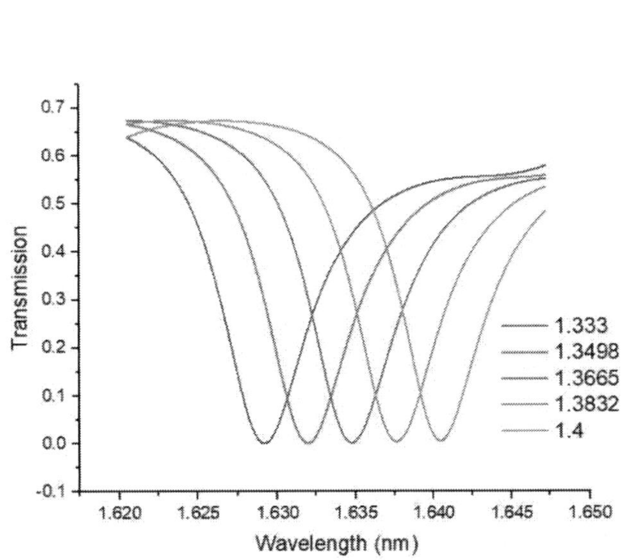

Figure 6. The effect of the surrounding refractive index change on the transmission spectrum.

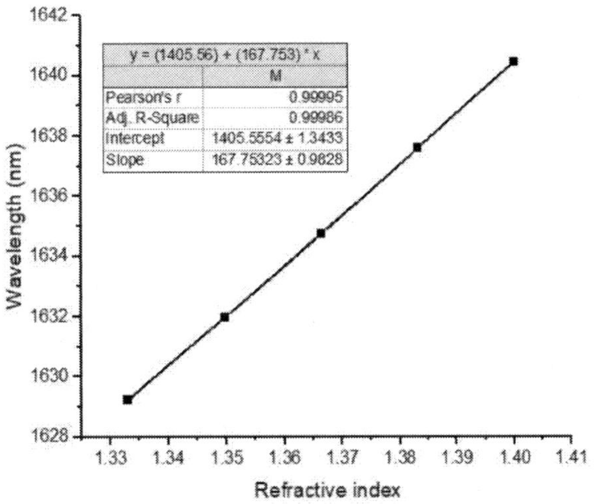

Figure 7: The model's response to refractive index change within a range of 1.333-1.400.

WGM resonators are sensitive to the change of surrounding refractive index and could be useful for sensing applications. The sensitivity of the proposed WGM resonator with the change of surrounding refractive index is comparable with the previous work on the gold-coated plasmonic WGM resonator [15,20,21]. WGM resonators are sensitive to the change of surrounding refractive index and could be useful for sensing applications. Table 1 compares the performance of this WGM resonator model with the previous WGM resonators, which were reported for various applications. The sensitivity of the proposed WGM resonator with the change of surrounding refractive index is comparable with the previous work on the gold-coated plasmonic WGM resonator.

Table 1 : Comparison of WGM sensors obtained based on different numerical analysis for various sensing applications.

Model	Sensitivity	Ref
WGM acoustic sensor	3.14 mV/Pa	[11]
Multi-layer WGM gyroscope	3.002 pm/(°.s^{-1})	[12]
Silicon WGM for dual measurements: **1. Refractive index** **2. Temperature**	45.8821 nm/RIU - 72.9402 nm/RIU 0.0730 nm/K - 0.0703 nm/K	[14]
Silicon WGM for dual measurements: **1. refractive index** **2. Pressure**	29.07 nm/RIU - 38.68 nm/RIU 0.576 pm/kPa - 0.589 pm/kPa	[19]
Gold-coated plasmonic WGM	180 nm/RIU	[15]
This model	167.8 nm/RIU	This work

4. Conclusion

MBRs confine light as a whispering gallery in the (r, □) plane. This work has investigated the WGM resonators' response to refractive index changes in the ranges 1.333 and 1.000. A COMSOL 2D model for the WGM resonator was used for refractive index detection. The 3 μm resonator showed a sensitivity of 167 nm/RIU in the refractive index range of 1.333-1.40 and showed a sensitivity of 95 nm/RIU in the refractive index range of 95 nm/RIU. The temperature effects on the resonator are shown to be 24 pm/°K in the temperature change range of 0-50 ΔK from room temperature. The sensing resonator has potential applications for biological and chemical analytes.

References

[1] Johari M A M, Apsari R and Yasin M, 2023/02/01 2023 "Comparative study an effect of polymethyl methacrylate coating on optical microbottle resonator for enhancing sensing performance for ethanol liquid sensor," *Microw. Opt. Technol. Lett.* **2** 690-96.

[2] Li H-C, Wang M-Y, Liu B, Liu J, Wang Q, He X-D, Ping Chan H, Wang D, Yuan J and Wu Q, 2022/06/15/ 2022 "Temperature-independent relative humidity sensing properties of polymer micro-bottle resonators coated with graphene oxide," *Measurement.* 111199.

[3] Zhu N, Shi B, Guo Y, Han B and Zhang Y-n, 2020 "Polydimethylsiloxane self-assembled whispering gallery mode microbottle resonator for ethanol sensing," *Opt. Mater.* 110024.

[4] Bhattacharya S, Veluthandath A V, Murugan G S and Bisht P B, 2020 "Temperature dependence of whispering gallery modes of quantum dot-doped microbottle resonators," *J. Lumin.*

[5] Baharuddin N A, Mokhtar N, Zain H A, Johari M A M, Apsari R and Harun S W, 2021/09/01/ 2021 "Micro-bottle resonator for sodium hypochlorite sensor," *Optik*. 167328.
[6] Bianucci P, Nov 2 2016 "Optical microbottle resonators for sensing," *Sensors (Basel)*. **11**
[7] Hou F, Zhang X, Wang Z, Yang L, Sun W, Yang Y, Dong Y, Huang Y and Wang T, 2020 "Magnetic fluid infiltrated microbottle resonator sensor with axial confined mode," *IEEE Photonics J.* **5** 1-9.
[8] Wang Z, Zhang X, Zhao S, Yu Y, Sun H, Yang Y, Dong Y, Huang Y and Wang T, 2021 "High-sensitivity flow rate sensor enabled by higher order modes of packaged microbottle resonator," *IEEE Photon. Technol. Lett.* **12** 599-602.
[9] Cai L, Pan J, Zhao Y, Wang J and Xiao S, 2020 "Whispering gallery mode optical microresonators: Structures and sensing applications," *physica status solidi*. **6** 1900825.
[10] Cai L, Pan J and Hu S, 2020 "Overview of the coupling methods used in whispering gallery mode resonator systems for sensing," *Opt Lasers Eng*.
[11] Cui J, Yu Y, Chu X, Zhao R, Zhu M, Zhang W and Zhang G, 2021 "Research on characteristics of broadband acoustic sensor based on silicon-based grooved microring resonator," *Micromachines*. **11** 1338.
[12] Xia D, Huang L and Zhao L, 2019 "A new design of an moems gyroscope based on a wgm microdisk resonator," *Sensors*. **12** 2798.
[13] Ippen E, Shank C and Dienes A, 1972 "Passive mode locking of the cw dye laser," *Applied Physics Letters*. **8** 348-50.
[14] Ma T, Yuan J, Sun L, Kang Z, Yan B, Sang X, Wang K, Wu Q, Liu H and Gao J, 2017 "Simultaneous measurement of the refractive index and temperature based on microdisk resonator with two whispering-gallery modes," *IEEE Photonics J.* **1** 1-13.
[15] Yin Y, Nie T and Ding M, 2020 "Refractive index sensor based on gold-coated whispering gallery mode microdisk resonator," *IEEE Sensors Journal*. **17** 9871-76.
[16] Yin Y, Nie T and Ding M, 2021 "Temperature sensor based on microbottle resonator immersed in isopropanol," *IEEE Photonics J.* **3** 1-7.
[17] Zhang Y-n, Wang M, Zhu N, Han B and Liu Y, 2023/01/15/ 2023 "Optical fiber hydrogen sensor based on self-assembled pdms/pd-wo3 microbottle resonator," *Sens. Actuators B Chem.* 132866.
[18] Foreman M R, Swaim J D and Vollmer F, 2015 "Whispering gallery mode sensors," *Advances in optics and photonics*. **2** 168-240.
[19] Zhang P, He D, Zhang C and Yan Z, 2020 "Fdtd simulation: Simultaneous measurement of the refractive index and the pressure using microdisk resonator with two whispering-gallery modes," *Sensors*. **14** 3955.
[20] Adnan Zain, H., Batumalay, M., Haris, H., Saad, I., Muhammad, A. R., Mustaffa, S. N., ... & Harun, S. W. (2023). Review of Microbottle Resonators for Sensing Applications. Micromachines, 14(4), 734.
[21] Tiu, Z. C., Harun, S. W., & Tan, S. J. (2023). Photon-to-photon Modulation in Fiber Laser System Using Emerging Materials. Optik, 171062.

Quantum Properties in Nonlinear Coupler with Raman Process.

Mohd Syafiq M. Hanapi[1], Abdel-Baset M. A. Ibrahim[1*]

[1]Faculty of Applied Sciences, Universiti Teknologi MARA (UiTM), Shah Alam, Selangor Malaysia
*Corresponding author's email: abdelbaset@uitm.edu.my

Abstract In this work, we have examined the generation of squeezed states of light in a two-waveguide coupler in which Raman processes are active in one waveguide. Both waveguides are mutually linearly interacting through the field's evanescent waves. We looked at a few interesting cases in which the system generated single mode squeezed states due to spontaneous and stimulated Raman processes. Under specific combinations of design parameters and phase mismatching conditions, squeezed states may take the form of collapses and revivals, or the second quadrature of the optical mode may oscillate completely below the short-noise limit of a coherent state.

Keywords; Quantum Optics; Nonlinear Coupler; Raman Process; Squeezes States

1. Introduction

In contrast to the conventional electromagnetic wave theory, in quantum theory, light is not seen as a wave but rather as a continuous stream of photons. How these photons interact with atoms and molecules is the focus of quantum optics, which aims to put to the test many of the counterintuitive predictions of quantum mechanics, such as squeezed states of light, entanglement, Photon antibunching, and other quantum phenomena. A new quantum technology based on these phenomena is emerging and requires experimentation through fundamental research. For instance, coherent light (usually from a laser source) in its classical form is hindered by an inherent standard noise limit, making it unsuited for use in cutting-edge quantum applications that require performance below the standard quantum limit. Squeezed light, however, exhibits reduced noise in one of the two electric field quadrature components [1]. This makes it suitable for use in high-precision measurements [2], such as the ultraprecise measurement of lengths for the detection of gravitational waves. Other potential applications of squeezed light include quantum teleportation (required for quantum computers) [3] and, communication [4]. To generate squeezed light, several nonlinear systems have been proposed such as Bose-Einstein condensates [5], parametric down-conversion [6], and nonlinear couplers [7]. These systems rely on the interaction of coherent light with a nonlinear material to generate squeezed light via various processes.

Among these quantum systems, the nonlinear coupler has received a significant deal of interest due to its advantages as a simple-structured, experimentally realizable, and easily integrated system. Since the standard two-waveguide coupler has been proposed [8], several designs of nonlinear couplers have been suggested such as three-waveguide [9], four-waveguide nonlinear couplers [10], [11], cavity-assisted nonlinear couplers [12], [13], and nonlinear couplers with frequency mismatch [14]. Also, nonlinear couplers based on various nonlinear processes such as Kerr nonlinear effect [15]–[18], second harmonic generation [19], and four-wave mixing have been proposed [20].

One of the most promising processes for producing nonclassical light is the Raman effect. Raman process is the third-order nonlinear scattering process that occurs when light interacts with solid material [21]. In this interaction, light transfers some of its energy to the solid, and this results in two possibilities: Stokes Raman and anti-Stokes Raman scattering (See Fig. 1a). Stokes Raman scattering occurs when the emitted light has lower energy than the incident light. Inversely, anti-Stokes Raman scattering takes place when the energy of the emitted light is higher than the incident light. In addition, both of these processes will cause phonon vibrations in solids. The quantum description of various forms of Raman process in the nonlinear media has been a topic of interest among quantum opticians (see, e.g., [22]). However, the quantum aspects of a nonlinear coupler operating with the Raman process are yet to be fully investigated [23], [24]. Particularly, we believe that the single and compound mode squeezing in the current system is yet to be investigated.

Here we consider a two-waveguide coupler system with one nonlinear Raman-active waveguide coupled to another waveguide with just a linear process. Each waveguide is pumped with one fundamental coherent mode. The coherent fundamental mode pumped into the Raman-active waveguide produces Raman Stokes mode, anti-Stokes mode, and phonon mode (See Fig. 1b). The fundamental modes in both waveguides are also linearly

Content from this work may be used under the terms of the Creative Commons Attribution 3.0 licence. Any further distribution of this work must maintain attribution to the author(s) and the title of the work, journal citation and DOI.
Published under licence by IOP Publishing Ltd

coupled (cross-action coupling) via evanescent waves. We will examine the generation of single and compound mode squeezed states in the current system due to spontaneous Raman under phase mismatching conditions.

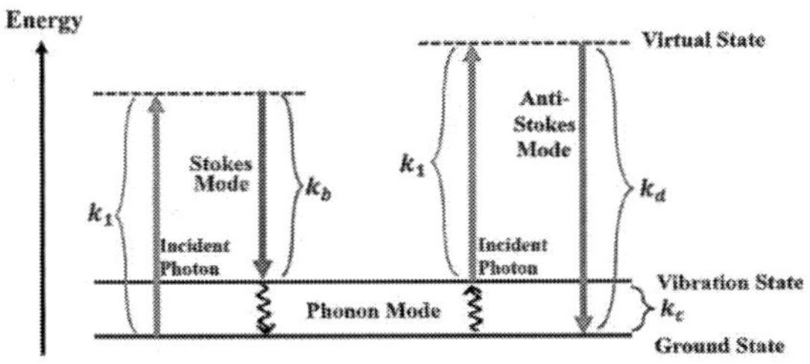

Figure 1a. Diagram for Raman process

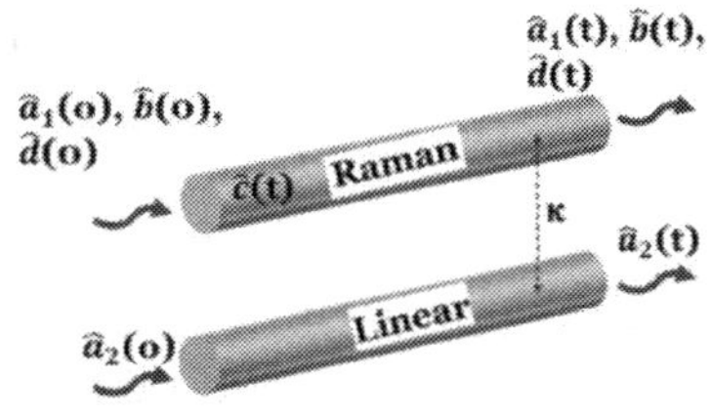

Figure 1b: Basic diagram of the two-waveguide asymmetric nonlinear coupler

2. Mathematical Description of the System

Here we use the analytical perturbative (AP) method to study the current system. The AP method was introduced by Sen and Mandal [25] and subsequently used by others to investigate the quantum optical properties of various quantum systems (See, e.g. [26]). The AP method has shown improved results compared with the usual short-length approximation (SLA) method used earlier in the literature (See, e.g., [27]). The quantum mechanical description of the current system can be expressed in terms of the following momentum operator.

$$\hat{G} = \hbar k_1 \hat{a}_1^\dagger \hat{a}_1 + \hbar k_2 \hat{a}_2^\dagger \hat{a}_2 + \hbar k_b \hat{b}^\dagger \hat{b} + \hbar k_c \hat{c}^\dagger \hat{c} + \hbar k_d \hat{d}^\dagger \hat{d} + \left(\hbar g \hat{a}_1 \hat{b}^\dagger \hat{c}^\dagger + \hbar \chi \hat{a}_1 \hat{c} \hat{d}^\dagger + \hbar \kappa \hat{a}_2 \hat{a}_1^\dagger + h.c.\right) \tag{1}$$

On the right side of equation (1), *h.c.* means the Hermitian conjugate, and the first two terms $(\hbar k_1 \hat{a}_1{}^\dagger \hat{a}_1, \hbar k_2 \hat{a}_2{}^\dagger \hat{a}_2)$ account for the pump modes in the first and second waveguides respectively. The third, fourth, and fifth terms $(\hbar k_b \hat{b}^\dagger \hat{b}, \hbar k_c \hat{c}^\dagger \hat{c}, \hbar k_d \hat{d}^\dagger \hat{d})$ accounts for the Stokes mode, the phonon mode, and the anti-Stokes mode, respectively. The coefficients k_j are the wave vectors. The term $\hbar g \hat{a}_1 \hat{b}^\dagger \hat{c}^\dagger + h.c.$ is responsible for the Stokes Raman scattering, where the annihilation of the pump mode creates the Stokes mode alongside the phonon mode. The coefficient g is the Stokes nonlinear coefficient that controls this process. Inversely, the term $\hbar \chi \hat{a}_1 \hat{c} \hat{d}^\dagger + h.c.$ indicates the anti-Stokes Raman scattering that creates the anti-Raman mode due to the annihilation of pump mode and phonon mode; the controlling parameter for this process is χ (The anti-Stokes nonlinear coefficient). Finally, the last term in Equation (1) $(\hbar \kappa \hat{a}_2 \hat{a}_1{}^\dagger + h.c.)$ expresses the evanescent linear coupling between the fundamental modes. The strength of this interaction is proportional to the evanescent coupling coefficient (κ); this coefficient is inversely proportional to the separation between the waveguides.

Using the Heisenberg picture, the evolution of an operator $\hat{a}_j$ be described through the spatial Heisenberg equation of motion in the form of $\frac{d\hat{a}_j(z)}{dz} = \frac{i}{\hbar}\left[\hat{a}_j(z), \hat{G}(z)\right]$ where z is a spatial variable along the propagation direction [28]. By substituting the momentum operator from Eq. (1) into the Heisenberg equation of motion for each one of the five operators $\hat{a}_j = \{\hat{a}_1, \hat{a}_2, \hat{b}, \hat{c}, \hat{d}\}$, we arrive at the following system differential equation for the operator's evolution.

$$\frac{d\hat{a}_1(z)}{dz} = ik_1\hat{a}_1(z) + ig\hat{b}(z)\hat{c}(z) + i\chi\hat{c}^\dagger(z)\hat{d}(z) + i\kappa\hat{a}_2(z) \tag{2}$$

$$\frac{d\hat{a}_2(z)}{dz} = ik_2\hat{a}_2(z) + i\kappa\hat{a}_1(z) \tag{3}$$

$$\frac{d\hat{b}(z)}{dz} = ik_b\hat{b}(z) + ig\hat{a}_1(z)\hat{c}^\dagger(z) \tag{4}$$

$$\frac{d\hat{c}(z)}{dz} = ik_c\hat{c}(z) + ig\hat{a}_1(z)\hat{b}^\dagger(z) + i\chi\hat{a}_1^\dagger(z)\hat{d}(z) \tag{5}$$

$$\frac{d\hat{d}(z)}{dz} = ik_d\hat{d}(z) + i\chi\hat{a}_1(z)\hat{c}(z) \tag{6}$$

These are the Heisenberg equations of motion that specifically describe the dynamics of each operator. In the analytical method, the solution to each operator in the previous system (2)-(6) is assumed in the form of the Baker-Hausdorff (BH) formula as below.

$$\hat{a}_j(z) = \exp\left[\frac{iz\hat{G}(z)}{\hbar}\right]\hat{a}_j(0)\exp\left[\frac{-iz\hat{G}(z)}{\hbar}\right] = \hat{a}_j(0) + \frac{iz}{\hbar}\left[\hat{G}(z),\hat{a}_j(0)\right] - \frac{1}{2}\frac{z^2}{\hbar^2}\left[\hat{G}(z),\left[\hat{G}(z),\hat{a}_j(0)\right]\right] + \dots \tag{7}$$

In the previous BH formula, $\hat{a}_j(0)$ represents the operator at its initial state, i.e., at $z = 0$. The expansion is usually limited to only second-order terms on the assumption that the higher-order terms are insignificant and can be disregarded. Upon evaluating the commutator terms $[\hat{G}(z), \hat{a}_j(0)]$ and $[\hat{G}(z), [\hat{G}(z), \hat{a}_j(0)]]$ in equation (7), closed-form analytical solutions to each operator could be obtained as follows.

$$\begin{aligned}\hat{a}_1(z) = {} & A_1(z)\hat{a}_1(0) + A_2(z)\hat{a}_2(0) + A_3(z)\hat{b}(0)\hat{c}(0) + A_4(z)\hat{c}^\dagger(0)\hat{d}(0) + A_5(z)\hat{a}_1(0)\hat{c}^\dagger(0)\hat{c}(0) \\ & + A_6(z)\hat{a}_1(0)\hat{b}(0)\hat{b}^\dagger(0) + A_7(z)\hat{a}_1^\dagger(0)\hat{b}(0)\hat{d}(0) + A_8(z)\hat{a}_1(0)\hat{d}^\dagger(0)\hat{d}(0)\end{aligned} \tag{8}$$

$$\hat{a}_2(z) = B_1(z)\hat{a}_2(0) + B_2(z)\hat{a}_1(0) + B_3(z)\hat{b}(0)\hat{c}(0) + B_4(z)\hat{c}^\dagger(0)\hat{d}(0) \tag{9}$$

$$\begin{aligned}\hat{b}(z) = {} & P_1(z)\hat{b}(0) + P_2(z)\hat{a}_1(0)\hat{c}^\dagger(0) + P_3(z)\hat{a}_2(0)\hat{c}^\dagger(0) + P_4(z)\hat{b}(0)\hat{c}(0)\hat{c}^\dagger(0) + P_5(z)\hat{c}^{\dagger 2}(0)\hat{d}(0) \\ & + P_6(z)\hat{a}_1(0)\hat{a}_1^\dagger(0)\hat{b}(0) + P_7(z)\hat{a}_1^2(0)\hat{d}^\dagger(0)\end{aligned} \tag{10}$$

$$\begin{aligned}\hat{c}(z) = {} & Q_1(z)\hat{c}(0) + Q_2(z)\hat{a}_1(0)\hat{b}^\dagger(0) + Q_3(z)\hat{a}_1^\dagger(0)\hat{d}(0) + Q_4(z)\hat{a}_2(0)\hat{b}^\dagger(0) + Q_5(z)\hat{a}_2^\dagger(0)\hat{d}(0) \\ & + Q_6(z)\hat{b}(0)\hat{b}^\dagger(0)\hat{c}(0) + Q_7(z)\hat{b}^\dagger(0)\hat{c}^\dagger(0)\hat{d}(0) + Q_8(z)\hat{a}_1(0)\hat{a}_1^\dagger(0)\hat{c} + Q_9(z)\hat{c}(0)\hat{d}^\dagger(0)\hat{d}(0) \\ & + Q_{10}(z)\hat{a}_1^\dagger(0)\hat{a}_1(0)\hat{c}(0)\end{aligned} \tag{11}$$

$$\begin{aligned}\hat{d}(z) = {} & R_1(z)\hat{d}(0) + R_2(z)\hat{a}_1(0)\hat{c}(0) + R_3(z)\hat{a}_2(0)\hat{c}(0) + R_4(z)\hat{b}(0)\hat{c}^2(0) + R_5(z)\hat{c}^\dagger(0)\hat{c}(0)\hat{d}(0) \\ & + R_6(z)\hat{a}_1^2(0)\hat{b}^\dagger(0) + R_7(z)\hat{a}_1(0)\hat{a}_1^\dagger(0)\hat{d}(0)\end{aligned} \tag{12}$$

Solutions (8)-(12) describe the spatial evolution of the operators $\hat{a}_1(z), \hat{a}_2(z), \hat{b}(z), \hat{c}(z)$ and $\hat{d}(z)$ in the z-direction expressed in terms of unknown spatial-dependent coefficients $A_k(z)$, $B_k(z)$, $P_k(z)$, $Q_k(z)$ and $R_k(z)$. These coefficients can be evaluated by substituting the operator solutions (8)-(12) into each equation of motion in the set (2)-(6) and collecting similar terms from both sides. This procedure produces a set of differential equations that define the spatial propagation of these unknown coefficients. This set is listed in Appendix A for completeness. As a result, even if the operator solutions in (8)-(12) have an analytical closed form, we still need to numerically solve the set of equations (A1) -(A36) to acquire their coefficients.

3. Single-Mode Squeezing

In quantum optics, a quantized coherent electric field $\hat{E}(t)$ is made up of two quadrature components with equally shared noise where $\hat{E}(t) = 2\lambda[\hat{X}\sin(\omega t - \vec{k}.\vec{r}) - \hat{Y}\cos(\omega t - \vec{k}.\vec{r})]$ and $\hat{X}, \hat{Y}$ are the amplitudes of these two quadrature components. In terms of creation ($\hat{a}$) and annihilation ($\hat{a}^\dagger$) operators, $\hat{X}, \hat{Y}$ are written as $\hat{X} = \frac{1}{2}(\hat{a} + \hat{a}^\dagger) and\ \hat{Y} = \frac{1}{2i}(\hat{a} - \hat{a}^\dagger)$ [29]. Squeezing is the condition, where the noise in one of the quadrature components is reduced below the shot noise level (The noise level of coherent state) at the expense of the increased noise in the other quadrature. Meaning that, if the first quadrature component is squeezed, the second quadrature component is not, and vice versa. Mathematically, for a specific single-mode j to be squeezed, one of the quadratures has to fulfill one of these conditions.

$$\left\langle\left(\Delta\hat{X}_j\right)^2\right\rangle=\left\langle\hat{X}_j^2\right\rangle-\left\langle\hat{X}_j\right\rangle^2<\frac{1}{4} \quad \text{or} \quad \left\langle\left(\Delta\hat{Y}_j\right)^2\right\rangle=\left\langle\hat{Y}_j^2\right\rangle-\left\langle\hat{Y}_j\right\rangle^2<\frac{1}{4} \tag{13}$$

where $\left\langle(\Delta\hat{X}_j)^2\right\rangle$ and $\left\langle(\Delta\hat{Y}_j)^2\right\rangle$ are the variances of the first and the second quadrature respectively. The value '1/4' in the previous equation is the standard noise level in a coherent state (shot noise level), and single-mode squeezing occurs when the noise in a certain mode j reduces below this value. The noise fluctuation obeys the Heisenberg uncertainty relation which states that it is impossible for both conditions in equation (13) to hold at the same time, i.e., $\sqrt{\langle(\Delta\hat{X}_j)^2\rangle\langle(\Delta\hat{Y}_j)^2\rangle} \geq 0.25$. In terms of creation and annihilation operators, the quadrature variances of some mode j can be written as.

$$\begin{bmatrix}\left\langle(\Delta\hat{X}_j)^2\right\rangle \\ \left\langle(\Delta\hat{Y}_j)^2\right\rangle\end{bmatrix}=\frac{1}{4}\left\{1+2\left\langle\hat{a}_j^{\dagger}\hat{a}_j\right\rangle-2\left\langle\hat{a}_j^{\dagger}\right\rangle\left\langle\hat{a}_j\right\rangle\pm\left[\left\langle\hat{a}_j^2\right\rangle-\left\langle\hat{a}_j\right\rangle^2+\left\langle\hat{a}_j^{\dagger 2}\right\rangle-\left\langle\hat{a}_j^{\dagger}\right\rangle^2\right]\right\} \tag{14}$$

In the previous equation (14), the bracket $\langle\ \rangle$ indicates the quantum expectation value of the operators in normal order. To obtain the quadrature variances for the first pump mode, we substitute $\hat{a}_1$ and its complex conjugate $\hat{a}_1^{\dagger}$ from equation (8) into equation (14). This yields the following equation.

$$\begin{aligned}\begin{Bmatrix}\left\langle\left(\Delta\hat{X}_1\right)^2\right\rangle \\ \left\langle\left(\Delta\hat{Y}_1\right)^2\right\rangle\end{Bmatrix}=\frac{1}{4}\Big\{1+2\Big[&|\alpha_d|^2|A_4(z)|^2+|\alpha_1|^2A_6(z)A_1^*(z)+\alpha_2^*\alpha_1A_6(z)A_2^*(z)+|\alpha_1|^2A_6(z)A_1(z)+\alpha_2\alpha_1^*A_6(z)A_2(z)\Big] \\ &\pm\Big[2\alpha_1\alpha_1A_6(z)A_1(z)+2\alpha_2\alpha_1A_6(z)A_2(z)+\alpha_b\alpha_dA_7(z)A_1(z)+\alpha_b\alpha_dA_4(z)A_3(z)+2\alpha_1^*\alpha_1^*A_6^*(z)A_1^*(z) \\ &+2\alpha_2^*\alpha_1^*A_6^*(z)A_2^*(z)+\alpha_b^*\alpha_d^*A_7^*(z)A_1^*(z)+\alpha_b^*\alpha_d^*A_4^*(z)A_3^*(z)\Big]\Big\}\end{aligned} \tag{15}$$

Similarly, we can develop an equation for possible squeezing exhibited by other modes. For example, if we substitute $\hat{a}_2$ and its complex conjugate $\hat{a}_2^{\dagger}$ from equation (9) into equation (14), we arrive at an expression for the quadrature variances for the second mode. However, here we will only look at squeezed states exhibited by the first pump mode, i.e., the mode propagates inside the Raman waveguide. We anticipate that the best possible squeezed states will be displayed by such a mode. In the previous equation, the variables α_j represent the initial complex amplitude of the operator $\hat{a}_j$, i.e., $\langle\hat{a}_j(z=0)\rangle=\alpha_j$ and $\langle\hat{a}_j^{\dagger}(z=0)\rangle=\alpha^*$ where $\alpha^*$is the complex conjugate of α.

4. Results and Discussions

It is easier to obtain numerical solutions by putting the relevant equations in dimensionless form. The system of equations (A1) -(A36) can be converted into dimensionless form by dividing both sides of each equation with k_1. The numerical procedure involves two main steps – viz. (i) solving simultaneously the coupled sets of Eqs. (A1)–(A36) to obtain numerical values of the spatial-dependent coefficients $\{A_k(z)\}_1^8$, $\{B_k(z)\}_1^4$, $\{P_k(z)\}_1^7$, $\{Q_k(z)\}_1^{10}$, and $\{R_k(z)\}_1^7$, and (ii) with the knowledge of these coefficients, the optical field quadrature variances for the first pumping mode $\langle(\Delta\hat{X}_j)^2\rangle, \langle(\Delta\hat{Y}_j)^2\rangle$ can be evaluated using Eq. (15).

In this section, numerical simulation results are presented. Here, only the result for the first pump mode in Raman-active waveguide will be shown. Under the spontaneous Raman process, the pump modes are prepared to be in the coherent state, while other modes are in the vacuum states, i.e., $\alpha_1 \neq 0,\ \alpha_2 \neq 0$ and $\alpha_b = \alpha_c = \alpha_d = 0$ [25]. The evolution of single-mode quadrature variances in the case of the spontaneous Raman process is shown in Figure 2. Observe the oscillation of both quadrature variances of the first pump mode below the shot noise level (y = ¼ line) of the coherent state. In particular, at certain distances, the second quadrature ΔY evolves totally below the shot-noise level. Here, both quadrature variances oscillate with 180° out of phase, which means, when the first quadrature variance reaches the maximum point, the second quadrature variance will be at its minimum point, and vice versa. In addition, the maximal squeezing amplitude of the second quadrature component ΔY is always lower than the first quadrature component ΔX, and this makes sense could be inferred from Equation (15).

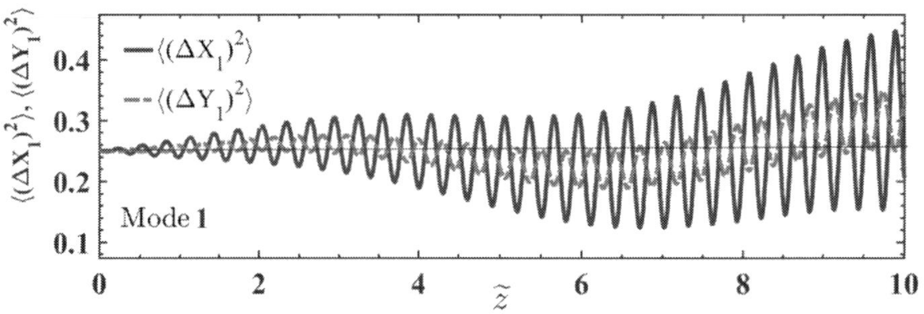

Figure 2: Evolution of the quadrature variances of the electric field of the pump mode $\hat{a}_1$ as a function of spatial distance showing fluctuation of squeezed states below the shot-noise level (Thin black line- $y = 0.25$) under spontaneous Raman process, i.e., when the pump modes are prepared in a coherent state, while other modes are in vacuum states, i.e., $\alpha_1 \neq 0$, $\alpha_2 \neq 0$ and $\alpha_b = \alpha_c = \alpha_d = 0$. Other parameters used in this simulation are fixed at $\tilde{k}_1 = \tilde{k}_2 = 10$, $\tilde{k}_b = 10$, $\tilde{k}_c = 0.00001$, $\tilde{k}_d = 10$, $\tilde{g} = \tilde{\chi} = 0.0001$ and $\tilde{\kappa} = 0.8$.

For the stimulated Raman process, all modes are initially set in coherent states [25]. In Figure 3, the graphs show the evolution of squeezed states as a function of scaled distance phase mismatching, i.e., when the wave numbers of both pumping modes are not equal, $\tilde{k}_1 \neq \tilde{k}_2$, which means their frequencies are not equal. In Fig. 3a, we choose the input pump fields to have scaled wavenumbers of $\tilde{k}_1 = 10$ and $\tilde{k}_2 = 12$ respectively. Under this condition, the squeezing in the first pump mode shows a collapses and revivals pattern. This behavior may result from the superposition of two modes with distinct phases. In this case, the second quadrature component ΔY oscillates almost below the shot noise level of a coherent state, which is an intriguing property. This is an excellent level of squeezed state with minimal noise. When the frequency discrepancy between the two pumping modes increases, as depicted in Fig. 3b, the collapses and revivals pattern begin to vanish and the quadrature variances of the first pump mode evolve with irregular amplitudes. As the mismatching increases further (Figure 3c), the amplitude of both quadrature variances increases progressively as a function of distance. This does not seem reasonable, and we anticipate that the AP method may produce inaccurate results for certain combinations of high initial input parameters and greater evolution distances. However, the AP method performs well with comparatively low initial input parameters, especially for early evolution distances.

We have finally examined the effect of linear coupling between the two pump modes on the generated squeezed states. Results are shown in Fig. 4a and Figure 4.b. for the case of linear coupling $\tilde{\kappa} = 0.2$, $\tilde{\kappa} = 0.7$ respectively. Figure 4a also depicts the evolution of the quadrature variances as a function of scaled distance. The oscillatory nature of the quadrature variances below the shot-noise level is evident. However, as observed in Fig.3c, the amplitude of quadrature variances steadily increases which seems unreasonable. Nevertheless, the general feature of the oscillatory nature of the quadrature variances around the shot-noise limit remains valid, especially at short evolution distances. An intriguing aspect of Figure 4b is that the first quadrature ΔX does not oscillate symmetrically about the shot-noise level and never oscillates completely above it. Nonetheless, at some distance ranges, the second quadrature variance ΔY oscillates above the shot noise level, indicating the absence of squeezing, whereas at other distance ranges, it oscillates completely below the shot noise level, indicating complete squeezing. In this case, this type of squeezing is only appropriate for short-distance signal transmission and not for long-distance optical communications.

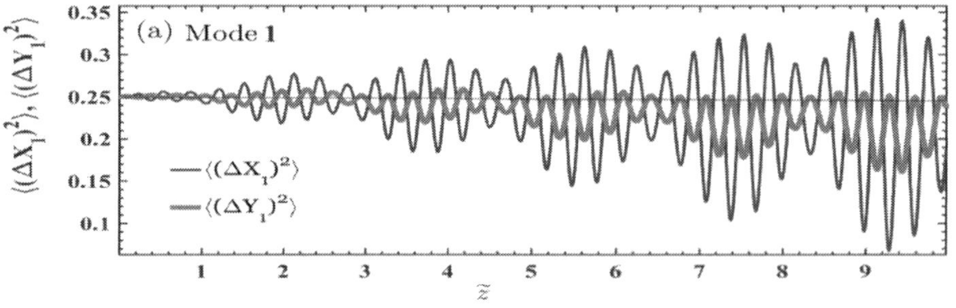

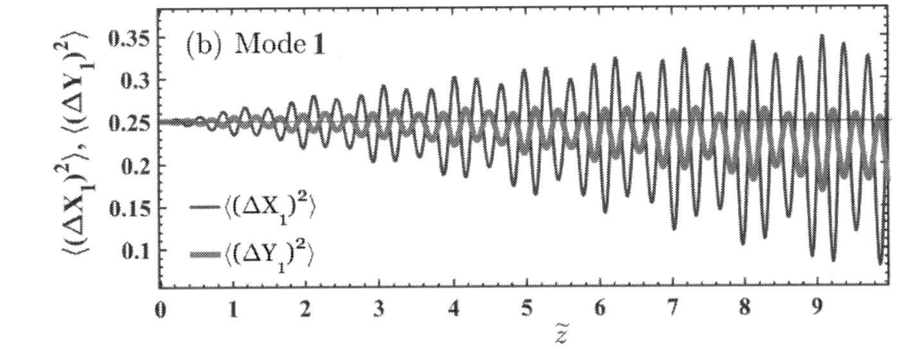

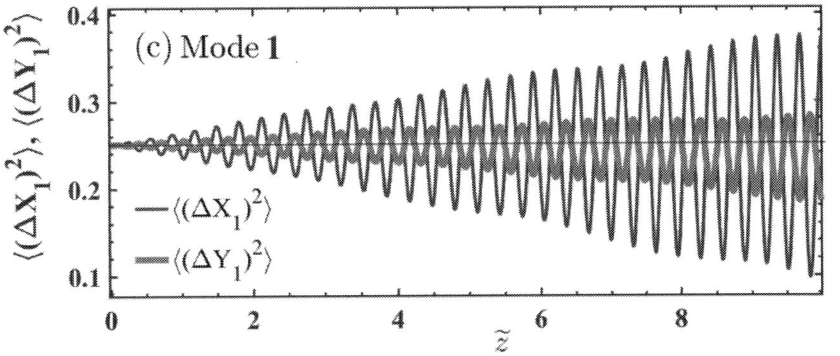

- Figure 3: Evolution of the quadrature variances of the electric field of the pump mode $\hat{a}_1$ as a function of spatial distance at different mismatching between the two pump modes showing fluctuation of squeezed states below the shot-noise level (Thin black line- $y = 0.25$) under stimulated Raman process, i.e., when all the modes are prepared in a coherent state with amplitudes $\alpha_1 = \alpha_2 = 10, \alpha_b = 9, \alpha_c = 0.01$ and $\alpha_d = 9$. a) under phase mismatching between pump modes and with $\tilde{k}_1 = 10$ and $\tilde{k}_2 = 13$. b) mismatching with $\tilde{k}_1 = 10$ and $\tilde{k}_2 = 16$. c) mismatching with $\tilde{k}_1 = 10$ and $\tilde{k}_2 = 28$. Other parameters remain as in Fig.2

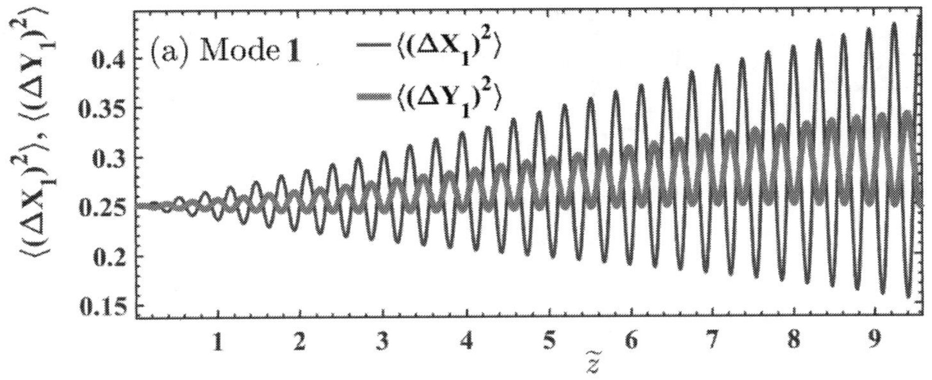

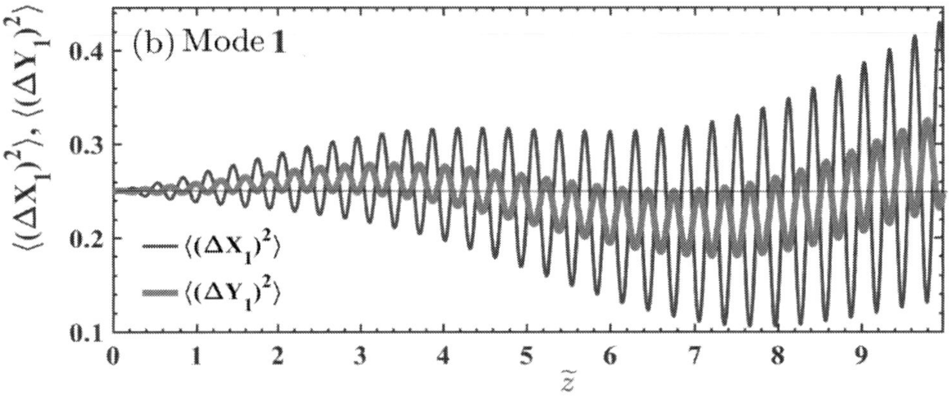

Figure 4: Evolution of the quadrature variances of the electric field of the pump mode $\hat{a}_1$ as a function of spatial distance at different values of linear coupling. a) $\tilde{\kappa} = 0.2$. b) $\tilde{\kappa} = 0.7$. The figures are generated under a stimulated Raman process with amplitudes $\alpha_1 = \alpha_2 = 10, \alpha_b = 9, \alpha_c = 0.01$ and $\alpha_d = 9$. Other parameters are $\tilde{k}_1 = \tilde{k}_2 = 10, \tilde{k}_b = 10, \tilde{k}_c = 0.00001, \tilde{k}_d = 10, \tilde{g} = \tilde{\chi} = 0.0001$.

5. Conclusion

To conclude, the AP method was used to study the generation of squeezed states in an asymmetric nonlinear coupler comprised of a Raman-active waveguide and a linear waveguide. In the AP method, analytical solutions of the operators are obtained assuming weak nonlinearity and expressions of quadrature variances of single mode squeezing can be derived. we have investigated the production of single-mode squeezed states of light in the Raman-active waveguide in cases of spontaneous and stimulated Raman processes. spontaneous Raman requires preparing the pump modes in coherent states and the Raman modes in vacuum states, while stimulated Raman requires preparing all modes in coherent states. during the spontaneous Raman process, the quadrature variances of the first waveguide mode oscillate around the shot noise level of the coherent state. The second quadrature evolves completely below the shot-noise level at specific distances. In the stimulated Raman process, all modes are initially set in coherent states. Under reasonable phase mismatching, the field shows squeezed states in the form of collapses and revivals patterns. This behavior may result from the superposition of two modes with distinct phases. In this case, the second quadrature component oscillates almost below the shot noise level of a coherent state as well. This is an excellent level of squeezed state with minimal noise. When the frequency mismatching between the two pumping modes increases, the collapses and revivals pattern begin to vanish and the quadrature variances of the first pump mode evolve with irregular amplitudes. We have finally examined the effect of linear coupling between the two pump modes on the generated squeezed states. Both field quadratures exhibit squeezed states of oscillatory nature with asymmetry about the shot-noise level. Nonetheless, at some distance ranges, the second quadrature variance oscillates above the shot noise level, indicating the absence of squeezing, whereas at other distance ranges, it oscillates completely below the shot noise level, indicating complete squeezing. In this case, this type of squeezing is appropriate for short-distance signal transmission in optical communications. This calculation paves the way for the future use of squeezed states in optical communications and precise measurements.

Acknowledgments
Abdel-Baset M. A. Ibrahim gratefully recognizes the Malaysia Ministry of Higher Education (MOHE) for the national grant FRGS/1/2021/STG07/UITM/02/10.

Appendix A

Equations (A1)–(A36) represents the evolution of the spatial-dependent coefficients $\{A_k(z)\}_1^8$, $\{B_k(z)\}_1^4$, $\{P_k(z)\}_1^7$, $\{Q_k(z)\}_1^{10}$, and $\{R_k(z)\}_1^7$. A numerical solution of these coupled equations is necessary to obtain these coefficients for the operator solutions (8)-(12). Here we have solved them using the 'ode45' feature in MATLAB software using the initial condition $A_1(0) = B_1(0) = P_1(0) = Q_1(0) = R_1(0) = 1$ and $x_m(0) = 0$, while other initial conditions are set to zero. i.e., $\{B_k(0)\}_2^4 = \{P_k(0)\}_2^7 = \{Q_k(z)\}_2^{10}$=$\{R_k(z)\}_1^7$=0.

Set for operator $\hat{a}_1$		$dP_7/dz = ik_bP_7 + igA_1Q_3^*$	(A19)
$dA_1/dz = ik_1A_1 + i\kappa B_2$	(A1)	Ser for operator $\hat{c}$	
$dA_2/dz = ik_1A_2 + i\kappa B_1$	(A2)	$dQ_1/dz = ik_cQ_1$	(A20)
$dA_3/dz = ik_1A_3 + igP_1Q_1 + i\kappa B_3$	(A3)	$dQ_2/dz = ik_cQ_2 + igA_1P_1^*$	(A21)
$dA_4/dz = ik_1A_4 + i\chi Q_1^*R_1 + i\kappa B_4$	(A4)	$dQ_3/dz = ik_cQ_3 + i\chi A_1^*R_1$	(A22)
$dA_5/dz = ik_1A_5 + igP_2Q_1 + i\chi Q_1^*R_2$	(A5)	$dQ_4/dz = ik_cQ_4 + igA_2P_1^*$	(A23)
$dA_6/dz = ik_1A_6 + igP_1Q_2$	(A6)	$dQ_5/dz = ik_cQ_5 + i\chi A_2^*R_1$	(A24)
$dA_7/dz = ik_1A_7 + igP_1Q_3 + i\chi Q_2^*R_1$	(A7)	$dQ_6/dz = ik_cQ_6 + igA_3P_1^*$	(A25)
$dA_8/dz = ik_1A_8 + i\chi Q_3^*R_1$	(A8)	$dQ_7/dz = ik_cQ_7 + igA_4P_1^* + i\chi A_3^*R_1$	(A26)
Set for operator $\hat{a}_2$		$dQ_8/dz = ik_cQ_8 + igA_1P_2^*$	(A27)
$dB_1/dz = ik_2B_1 + i\kappa A_2$	(A9)	$dQ_9/dz = ik_cQ_9 + i\chi A_4^*R_1$	(A28)
$dB_2/dz = ik_2B_2 + i\kappa A_1$	(A10)	$dQ_{10}/dz = ik_cQ_{10} + i\chi A_1^*R_2$	(A29)
$dB_3/dz = ik_2B_3 + i\kappa A_3$	(A11)	Set for operator $\hat{d}$	
$dB_4/dz = ik_2B_4 + i\kappa A_4$	(A12)	$dR_1/dz = ik_dR_1$	(A30)
Set for operator $\hat{b}$		$dR_2/dz = ik_dR_2 + i\chi A_1Q_1$	(A31)
$dP_1/dz = ik_bP_1$	(A13)	$dR_3/dz = ik_dR_3 + i\chi A_2Q_1$	(A32)
$dP_2/dz = ik_bP_2 + igA_1Q_1^*$	(A14)	$dR_4/dz = ik_dR_4 + i\chi A_3Q_1$	(A33)
$dP_3/dz = ik_bP_3 + igA_2Q_1^*$	(A15)	$dR_5/dz = ik_dR_5 + i\chi A_4Q_1$	(A34)
$dP_4/dz = ik_bP_4 + igA_3Q_1^*$	(A16)	$dR_6/dz = ik_dR_6 + i\chi A_1Q_2$	(A35)
$dP_5/dz = ik_bP_5 + igA_4Q_1^*$	(A17)	$dR_7/dz = ik_dR_7 + i\chi A_1Q_3$	(A36)
$dP_6/dz = ik_bP_6 + igA_1Q_2^*$	(A18)		

REFERENCES

[1] D. F. Walls, "Squeezed states of light," *nature*, vol. 306, no. 5939, pp. 141–146, 1983.

[2] K. Liu, C. Cai, J. Li, L. Ma, H. Sun, and J. Gao, "Squeezing-enhanced rotating-angle measurement beyond the quantum limit," *Appl. Phys. Lett.*, vol. 113, no. 26, p. 261103, Dec. 2018, doi: 10.1063/1.5066028.

[3] C. Xin-Hua, "Quantum teleportation of entangled squeezed vacuum states," *Chin. Phys.*, vol. 12, no. 10, p. 1066, 2003.

[4] F. Kaiser, B. Fedrici, A. Zavatta, V. D'Auria, and S. Tanzilli, "A fully guided-wave squeezing experiment for fiber quantum networks," *Optica*, vol. 3, no. 4, p. 362, Apr. 2016, doi: 10.1364/OPTICA.3.000362.

[5] M. T. Johnsson, G. R. Dennis, and J. J. Hope, "Squeezing in Bose–Einstein condensates with large numbers of atoms," *New J. Phys.*, vol. 15, no. 12, p. 123024, Dec. 2013, doi: 10.1088/1367-2630/15/12/123024.

[6] L. Shiqun and W. Lingan, "Degenerate parametric down conversion of squeezed light," *Chin. Phys. Lett.*, vol. 8, no. 5, p. 222, 1991.

[7] A.-B. M. Ibrahim, B. A. Umarov, and M. R. B. Wahiddin, "Squeezing in the Kerr nonlinear coupler via phase-space representation," *Phys. Rev. A*, vol. 61, no. 4, p. 043804, 2000.

[8] S. Jensen, "The nonlinear coherent coupler," *IEEE J. Quantum Electron.*, vol. 18, no. 10, pp. 1580–1583, 1982.

[9] M. S. M. Hanapi, A.-B. M. Ibrahim, R. Julius, and P. K. Choudhury, "Squeezed Light Generation in Three-Channel Nonlinear Coupler with Second-Harmonic Generation and Frequency Mismatch," in *J. Phys. Conf. Ser.*, IOP Publishing, 2022, p. 012021.

[10] R. Julius, A.-B. M. A. Ibrahim, P. K. Choudhury, and H. Eleuch, "Quantum states generation in multichannel directional coupler with second-order nonlinearity," *Optik*, vol. 186, pp. 212–220, 2019.

[11] R. Julius, A.-B. M. A. Ibrahim, M. K. Abd-Rahman, and P. K. Choudhury, "Quantum dynamics of a four-channel Kerr nonlinear directional coupler," *Opt. Rev.*, vol. 25, no. 5, pp. 563–570, 2018.

[12] R. Julius, A.-B. M. A. Ibrahim, P. K. Choudhury, and H. Eleuch, "On the nonclassical dynamics of cavity-assisted four-channel nonlinear coupler," *Chin. Phys. B*, vol. 27, no. 11, Art. no. 11, Nov. 2018, doi: 10.1088/1674-1056/27/11/114206.

[13] R. Julius, A.-B. M. A. Ibrahim, and M. S. Deni, "Quantum state generation in a four-mode Kerr nonlinear directional coupler," *Laser Physics*, vol. 24, no. 3, p. 035202, 2014.

[14] R. Julius, M. Zahirzai, A.-B. M. A. Ibrahim, H. Eleuch, and P. K. Choudhury, "Single- and compound-mode squeezing in nonlinear coupler with frequency mismatch," *J. Electromagn. Waves Appl.*, pp. 1–15, Dec. 2019, doi: 10.1080/09205071.2019.1702591.

[15] Š. Bräuer and P. Marek, "Generation of quantum states with nonlinear squeezing by Kerr nonlinearity," *Opt. Express*, vol. 29, no. 14, pp. 22648–22658, 2021.

[16] M. S. M. Hanapi, A.-B. M. A. Ibrahim, R. Julius, and H. Eleuch, "The Quantum Kerr Nonlinear Coupler: The Analytical Versus Phase-Space Method," *Can. J. Phys.*, p. In press, 2021, doi: https://doi.org/10.1139/cjp-2020-0389.

[17] M. S. M. Hanapi, A.-B. M. A. Ibrahim, and P. K. Choudhury, "On the perturbative approach to analyse entanglement in two-channel Kerr nonlinear coupler," *Optik*, vol. 243, p. 167420, Oct. 2021, doi: 10.1016/j.ijleo.2021.167420.

[18] M. S. M. Hanapi, A.-B. M. Ibrahim, and R. Julius, "On the Detection of Single-Mode Squeezed States in Kerr Nonlinear Coupler," in *J. Phys. Conf. Ser.*, IOP Publishing, 2021, p. 012002.

[19] P. Dong, J. Upham, A. Jugessur, and A. G. Kirk, "Observation of continuous-wave second-harmonic generation in semiconductor waveguide directional couplers," *Opt. Express*, vol. 14, no. 6, pp. 2256–2262, 2006.

[20] A. Sheveleva, U. Andral, B. Kibler, P. Colman, J. M. Dudley, and C. Finot, "Idealized four-wave mixing dynamics in a nonlinear Schrödinger equation fiber system," *Optica*, vol. 9, no. 6, p. 656, Jun. 2022, doi: 10.1364/OPTICA.445172.

[21] D. W. Shipp, F. Sinjab, and I. Notingher, "Raman spectroscopy: techniques and applications in the life sciences," *Adv. Opt. Photon.*, vol. 9, no. 2, p. 315, Jun. 2017, doi: 10.1364/AOP.9.000315.

[22] K. Thapliyal, A. Pathak, B. Sen, and J. Peřina, "Lower-and higher-order nonclassical features in non-degenerate hyper-Raman processes," *Opt. Commun.*, vol. 444, pp. 111–119, 2019.

[23] M. Das, K. Thapliyal, B. Sen, J. Peřina, and A. Pathak, "Interplay between quantum Zeno and anti-Zeno effects in a nondegenerate hyper-Raman nonlinear optical coupler," *Phys. Rev. A*, vol. 103, no. 1, p. 013713, Jan. 2021, doi: 10.1103/PhysRevA.103.013713.

[24] J. Perina and J. Perina, "Statistics of light in Raman and Brillouin nonlinear couplers," *Quantum Semiclass. Opt.*, vol. 9, no. 3, pp. 443–464, Jun. 1997, doi: 10.1088/1355-5111/9/3/013.

[25] B. Sen and S. Mandal, "Squeezed states in spontaneous Raman and in stimulated Raman processes," *J. Mod. Opt.*, vol. 52, no. 13, Art. no. 13, Sep. 2005, doi: 10.1080/09500340500072984.

[26] N. Alam, K. Thapliyal, A. Pathak, B. Sen, A. Verma, and S. Mandal, "Bose-condensed optomechanical-like system and a Fabry–Perot cavity with one movable mirror: quantum correlations from the perspectives of quantum optics," *Eur. Phys. J. D*, vol. 73, pp. 1–12, 2019.

[27] J. Peřina and J. Bajer, "Non-classical Light in Nonlinear Symmetric and Asymmetric Couplers," *J. Mod. Opt.*, vol. 42, no. 11, pp. 2337–2346, Nov. 1995, doi: 10.1080/09500349514552031.

[28] J. Perina and J. Perina, "Quantum statistical properties of codirectional and contradirectional nonlinear couplers with phase mismatch," *Quantum Semiclass. Opt.*, vol. 7, no. 5, pp. 863–876, Oct. 1995, doi: 10.1088/1355-5111/7/5/008.

[29] Z. Ficek and M. R. Wahiddin, *Quantum Optics for Beginners*, 0 ed. Jenny Stanford Publishing, 2016. doi: 10.1201/b16875.

Sensitivity Performance of Optical Waveguide Sensors with Different Shapes of Cladding Layer for Microplastics Detection

A I M Hanafi[1], N N H Saris[1], S M Idrus[1] and A Hamzah[2]

[1] Faculty of Electrical Engineering, Universiti Teknologi Malaysia, 81310 UTM Johor Bahru, Johor Darul Ta'zim, Malaysia
[2] Department of Electronic Systems Engineering, Malaysia-Japan International Institute of Technology, Universiti Teknologi Malaysia, 54100, Kuala Lumpur, Malaysia

nurnajahatulhuda@utm.my

Abstract. With microplastics pollution becoming a global concern, there comes a need for sensors to attain an optimal level of sensitivity to detect microplastics in water. This work investigated the effects of cladding layer shapes on the sensitivity performance of an optical waveguide sensor for microplastics detection in water. In this research, three different cladding shapes—C-shaped fiber, D-shaped fiber, and rectangular waveguide with circular core—were simulated by using Wave Optics Module-COMSOL Multiphysics® software. The results indicated that the C-shaped fiber exhibited significantly higher sensitivity, with a sensitivity value of 1.070×10^{-3} compared to the D-shaped fiber and rectangular waveguide with 3.845×10^{-4} and 3.842×10^{-4}, respectively. The sensitivities of the D-shaped fiber and rectangular waveguide were relatively similar and did not exhibit any significant difference. The higher sensitivity of the C-shaped fiber is attributed to its larger exposed core area to the analyte, resulting in higher interaction of the evanescent wave with the analyte. However, fabricating the C-shaped fiber is more challenging compared to the other two shapes. This research highlights the significance of cladding shapes in optical waveguide sensor sensitivities and provides design optimization insights for microplastics detection in water.

1. Introduction

Plastics have numerous uses in various industries and are commonly utilized in day-to-day life such as in packaging, food containers, and drinking bottles. Nonetheless, improper handling of discarded plastics can result in ecological degradation, particularly in water habitats like rivers, oceans, and lakes [1]. This can lead to the formation of microplastic fragments that pose a risk to aquatic animals and human health [2].

Several studies have been conducted particularly in detecting microplastics in water by adopting methods such as Raman [3], near infrared (IR) [4], and Fourier transform infrared (FTIR) spectroscopy [5]. Nevertheless, some disputed issues have risen with respect to these methods, which include their complex processing steps, low yield, and high costs, thus making them time-consuming and inefficient [6-8].

Content from this work may be used under the terms of the Creative Commons Attribution 3.0 licence. Any further distribution of this work must maintain attribution to the author(s) and the title of the work, journal citation and DOI.
Published under licence by IOP Publishing Ltd

To address these limitations, a proposed solution was to simulate an optical waveguide sensor using Wave Optics Module-COMSOL Multiphysics® software. The proposed method offers prompt and in-situ detection, effectively mitigating the aforementioned issues [9]. Furthermore, three commonly known shapes of cladding waveguides were analyzed to enhance the sensor's sensitivity through further optimization. Furthermore, the study investigated the effective refractive index changes at different analyte refractive index mediums, along with the core principles of physics behind these factors.

2. Simulation Settings

The cross-sections of the novel europium-aluminum doped polymer composite optical planar waveguide sensor for the three different cladding shapes—C-shaped fiber, D-shaped fiber, and rectangular waveguide with circular core—are illustrated in **Figure 1(a)**, **Figure 1(b)**, and **Figure 1(c)**, respectively. The sensor comprises a core with diameter of 9 μm and refractive index of 1.510, surrounded by cladding with refractive index of 1.501. The operating wavelength of 617 nm was chosen because it has the lowest attenuation window of a practical waveguide [10]. The refractive index range for microplastics in water was set between 1.480 and 1.500 with an RI step increment of 0.005. In this range, 1.490 RIU represents the refractive index of low-density polyethylene (LDPE) [1].

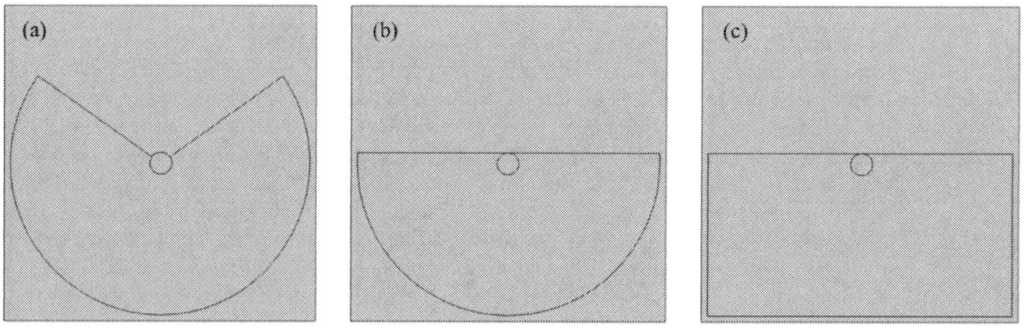

Figure 1 *Cross-sectional view of the optical waveguide sensor with three different cladding shapes: **(a)** C-shaped fiber, **(b)** D-shaped fiber and **(c)** rectangular waveguide with circular core*

3. Results and Discussion

The electric field distribution of all three waveguides with different cladding shapes can be analyzed by studying the mode field diameter (MFD) at a selective analyte refractive index (RI) of 1.490, as shown in **Figure 2**. The core-analyte border is located at the 4.5 μm position, allowing the penetration depth of the evanescent wave to be examined in the post-border region as shown in the inset of **Figure 2**. In this region, the C-shaped fiber exhibits a more significant penetration of the evanescent wave into the analyte compared to the other two cladding types. This increased penetration enables higher interaction between the evanescent wave and the analyte, resulting in higher sensitivity of the sensor.

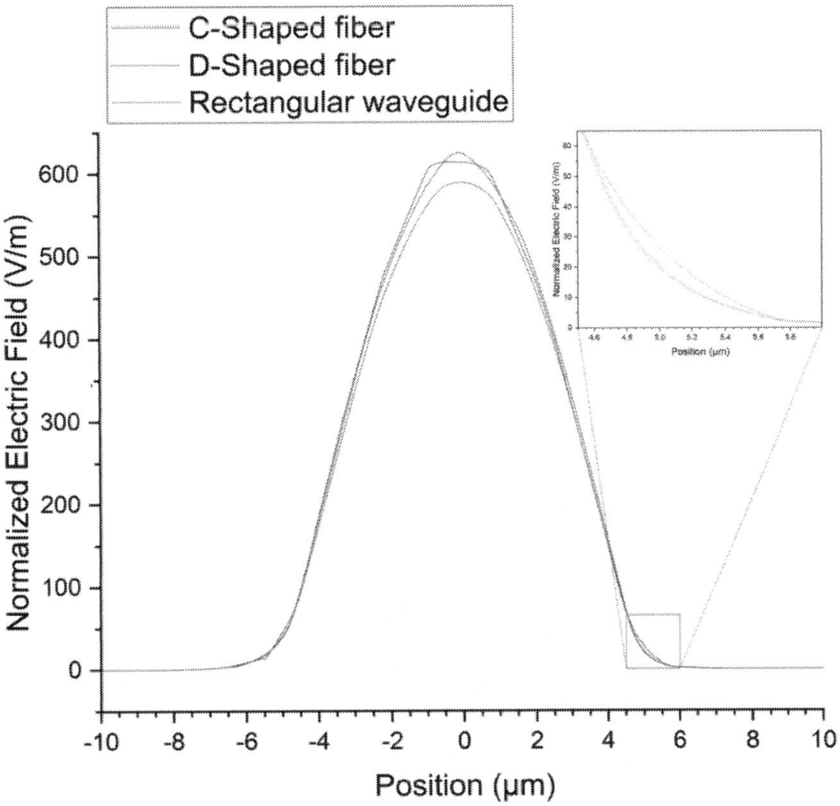

Figure 2 *MFD of the C-shaped fiber, D-shaped fiber, and rectangular waveguide at a selective analyte RI of 1.49*

Different effective mode index, n_{eff} can be achieved by varying the analyte RI. The changes in n_{eff}, referred to as Δn_{eff}, were plotted at different analyte RIs as shown in **Figure 3**. The inset graph in **Figure 3** provides a closer look at the D-shaped fiber and rectangular waveguide curves, which were separated when magnified. With an increase in the refractive index (RI) of the analyte from 1.4800 to 1.5000, the Δn_{eff} experiences a non-linear growth. This non-linear variation arises from the nonlinearity exhibited by the evanescent wave energy during its interaction with the analyte medium [10]. Nonetheless, Xiao et al. proposed a method for assessing sensitivity by examining the changes with a linear fit, thereby easily determining sensitivity based on the graph's slope [10]. As a result, the sensor recorded an estimated sensitivity for the C-shaped-fiber, D-shaped fiber, and rectangular waveguide as 1.070×10^{-3}, 3.845×10^{-4} and 3.842×10^{-4} respectively, in dimensionless unit.

The results of the study showed that the C-shaped fiber had a significantly higher sensitivity compared to the D-shaped fiber and rectangular waveguide. Conversely, no significant difference in sensitivity was observed between the D-shaped fiber and rectangular waveguide. This similarity is attributed to the indistinguishable cladding layer surrounding both the D-shaped fiber and rectangular waveguide from the core section's perspective. On the other hand, the increased sensitivity of the C-shaped fiber can be attributed to its larger exposed core area to the analyte, leading to a higher interaction of the evanescent wave with the analyte. Nonetheless, the production of the C-shaped fiber is relatively more complex compared to the other two shapes.

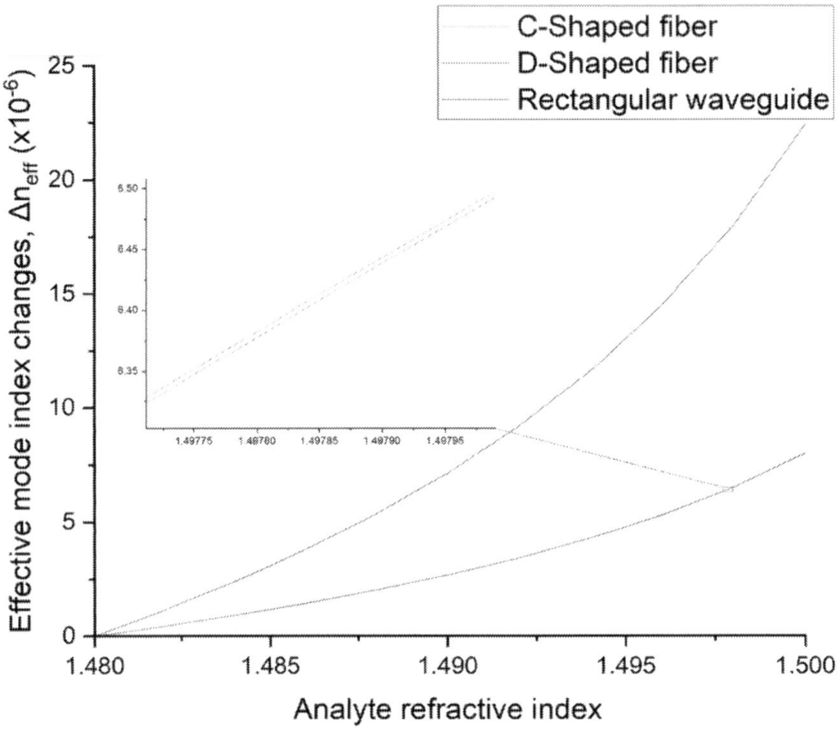

Figure 3 *Effective mode index changes Δn_{eff} versus analyte refractive index for all three cladding shapes: C-shaped fiber, D-shaped fiber, and rectangular waveguide. The inset graph provides a closer look at the D-shaped fiber and rectangular waveguide curves, which were separated when magnified.*

4. Conclusion

The study demonstrated the simulation of an optical waveguide sensor with different cladding shapes for detecting microplastics in water. The sensor was simulated for a range of analyte refractive indices from 1.480 to 1.500, relative to the microplastic refractive index. The C-shaped fiber exhibited the highest sensitivity of 1.070×10^{-3}, while the D-shaped fiber and rectangular waveguide had sensitivities of 3.845×10^{-4} and 3.842×10^{-4}, respectively, in dimensionless units. This higher sensitivity of the C-shaped fiber is attributed to its larger exposed core area to the analyte, which enhances the interaction of the evanescent wave with the analyte. Overall, the simulated optical waveguide sensor designs provide valuable insights into increasing the potential of optical waveguide sensors for microplastics detection in water.

Acknowledgement

This research is a collaborative project of Faculty of Electrical Engineering and Malaysia Japan International Institute of Technology (MJIIT), Universiti Teknologi Malaysia (UTM) and Keio University, funded by UTM Fundamental Research (UTMFR) grant vot number Q.J130000.3851.21H76 and Leave a Nest grant. The authors would like to thank the Takaaki Ishigure Informatics Laboratory, Keio University, and Lightwave Communication Research Group (LCRG) as well as Optical Device System Laboratory (ODESY) members for their help and encouragement.

References

1. Asamoah, B.O., et al., A prototype of a portable optical sensor for the detection of transparent and translucent microplastics in freshwater. Chemosphere, 2019. **231**: p. 161-167.
2. Yadav, H., S. Sethulekshmi, and A. Shriwastav, Estimation of microplastic exposure via the composite sampling of drinking water, respirable air, and cooked food from Mumbai, India. Environmental Research, 2022. **214**.
3. Fu, D., et al., Quantitative Chemical Imaging with Multiplex Stimulated Raman Scattering Microscopy. Journal of the American Chemical Society, 2012. **134**(8): p. 3623-3626.
4. Liu, J.B., Y.C. Du, and Y.H. Zhao, Soil Microplastics Spectrum Based on Visible near-Infrared Spectroscopy. Bangladesh Journal of Botany, 2022. **51**(4): p. 971-977.
5. Michel, A.P.M., et al., Comparison of Laboratory Analytical Techniques for the Rapid Identification of Marine Plastics. Oceans 2019 - Marseille, 2019.
6. Ainali, N.M., et al., Do poly(lactic acid) microplastics instigate a threat? A perception for their dynamic towards environmental pollution and toxicity. Science of the Total Environment, 2022. **832**.
7. Mohammed, F., et al., Analytical methods to detect adulteration of argan oil: A critical review. Microchemical Journal, 2021. **168**.
8. Mogha, N.K. and D. Shin, Nanoplastic detection with surface enhanced Raman spectroscopy: Present and future. Trac-Trends in Analytical Chemistry, 2023. **158**.
9. Razali, N.M. and N.N.H. Saris. Simulation of Optical Planar Waveguide Sensor for Microplastics Detection in Water. in Proceedings of the International Conference on Numerical Simulation of Optoelectronic Devices, NUSOD. 2022.
10. Saris, N.N.H., et al., Waveguide length and pump power effects on the amplification of europium aluminum doped polymer. Optik, 2021. **239**.

Investigating the Impact of SMF on Brillouin Fiber Laser Performance

H N A Ali[1,5], N Arsad[1*], N F Zulkipli[3], A H A Rosol[4], T N Abdali[5], H Q Ghadhban[6] and S W Harun[1*]

[1]Photonics Engineering Laboratory, Department of Electrical Engineering, Faculty of Engineering, University of Malaya, 50603 Kuala Lumpur, Malaysia.
[2]Department of Electrical, Electronic and Systems Engineering, Faculty of Engineering and Built Environment, Universiti Kebangsaan Malaysia, 43600 Bangi, Malaysia.
[3]Department of Engineering and Built Environment, Tunku Abdul Rahman University of Management and Technology (TAR UMT) Penang Branch, 11200 Tanjong Bungah, Pulau Pinang, Malaysia.
[4]Department of Electronic Systems Engineering, Malaysia–Japan International Institute of Technology, Universiti Teknologi Malaysia, Jalan Sultan Yahya Petra, 54100 Kuala Lumpur, Malaysia.
[5]Faculty of Basic Education, University of Misan, 62001, Iraq.
[6]Department of postgraduate center, university of diyala, 32001, Iraq.

* Corresponding author: noa@ukm.edu.my, swharun@um.edu.my

Abstract. Using erbium-doped fiber, a steady Q-switched erbium-doped laser (EDFL) was successfully demonstrated. With a 150 m of Single Mode Fiber to generate the Brillouin Fiber Laser and increase the non-linearity inside the cavity. In the absence of a Brillouin pump, a compact Brillouin fiber laser (BFL) is realized by combining an Erbium Doped Fiber (EDF) and a Single Mode Fiber (SMF), which serve as both the Brillouin and linear gain media. It is accomplished by self-Brillouin production in a laser cavity from 4.5 m of EDF and 150 m of SMF with strong nonlinearity. The SMF inside the ring cavity was used to generate self-generating BFL. The BFL in the linear cavity produced more Brillouin stokes at a lower pump threshold. In the experiments, the average of spacing is 0.088 nm was generated at the EDF pump power of 93.468 to 259.493 mW for the 150 m of SMF. This compact BFL has total cavity length (160.34 m). Eventually, this is the most recent gain medium length recorded for Brillouin erbium fiber laser production without the use of a Brillouin pump.

1. Introduction

A Brillouin fiber laser is a type of laser that utilizes the phenomenon of Brillouin scattering to generate coherent light. It operates by exploiting the interaction between a high-power pump laser and an optical fiber that contains a gain medium capable of exhibiting stimulated Brillouin scattering[1]. Brillouin scattering is a nonlinear optical effect that occurs when light propagates through a material and interacts with its acoustic vibrations[2]. It leads to the scattering of photons by acoustic phonons, resulting in the generation of a new optical wave with a slightly shifted frequency called the Brillouin scattered wave[3].

In a Brillouin fiber laser, a pump laser with a high-power level is coupled into an optical fiber[4]. The pump laser creates a strong optical intensity within the fiber, which induces acoustic vibrations due to the electrostriction effect[5]. These acoustic vibrations interact with the pump light, leading to Brillouin scattering[6]. This scattering process generates a backward-propagating Brillouin scattered wave with a frequency shift equal to the Brillouin frequency shift, which is dependent on the fiber's properties[7].

Content from this work may be used under the terms of the Creative Commons Attribution 3.0 licence. Any further distribution of this work must maintain attribution to the author(s) and the title of the work, journal citation and DOI.
Published under licence by IOP Publishing Ltd

The backward-propagating Brillouin scattered wave can be reflected back into the fiber by a fiber Bragg grating or other means[8]. As it propagates in the opposite direction of the pump laser, it experiences further amplification through the process of stimulated Brillouin scattering[9]. This feedback mechanism sustains the laser oscillation and allows the generation of coherent light at the Brillouin frequency[10].

Brillouin fiber lasers have unique properties, including narrow linewidth and low intensity noise, making them useful in various applications such as fiber optic sensing, optical communications, and microwave photonics[11].

To the best of our knowledge, add a single-mode fiber (SMF) to a Brillouin fiber laser, it can have several effects on the laser's performance:

Mode Conditioning: The SMF can act as a mode conditioning element for the laser. It helps to shape the output beam profile by modifying the transverse mode structure of the laser beam. This can be beneficial in applications where a specific beam profile or mode quality is required[11].

Increased Fiber Length: By adding an SMF, the overall length of the laser cavity is increased[12]. This can have an impact on the laser's performance, such as altering the threshold pump power required for lasing or affecting the laser's output power and efficiency. The longer fiber length may also introduce additional losses and dispersion, which can influence the laser's spectral characteristics[13].

Enhanced Brillouin Gain: The addition of an SMF can increase the Brillouin gain in the laser cavity[14]. The Brillouin gain is dependent on the fiber's properties, such as its length, composition, and effective area[15]. By introducing an SMF with specific properties, such as a longer length or higher Brillouin gain coefficient, the Brillouin scattering process can be enhanced, leading to higher output power and improved laser performance[16].

Increased Stability: The addition of an SMF can help stabilize the laser's operation[17]. It can provide a feedback mechanism to suppress unwanted cavity fluctuations and improve the laser's output stability[18]. The SMF can act as a distributed feedback element, reducing mode hopping and enhancing the laser's coherence properties[19].

Dispersion Effects: The SMF may introduce dispersion effects into the laser cavity. Dispersion refers to the dependence of the refractive index on the light's wavelength, and it can influence the laser's pulse duration, chirp, or spectral width. Depending on the specific characteristics of the SMF, such as its dispersion profile and length, these effects can be tailored to meet specific application requirements.

It is important to note that the specific impact of adding an SMF to a Brillouin fiber laser will depend on various factors, including the properties of the SMF, the design of the laser cavity, and the intended application. Careful consideration and optimization are required to achieve the desired performance enhancements or modifications[20].

In conclusion, the Brillouin fiber laser is a type of laser that utilizes the phenomenon of Brillouin scattering to generate coherent light. It operates by exploiting the interaction between a high-power pump laser and an optical fiber containing a gain medium capable of exhibiting stimulated Brillouin scattering. Brillouin fiber lasers have unique properties such as narrow linewidth and low intensity noise, making them useful in various applications[20].

When adding a single-mode fiber (SMF) to a Brillouin fiber laser, several effects can occur. The SMF can act as a mode conditioning element, shaping the output beam profile. It can increase the overall fiber length, which may impact the laser's performance, threshold pump power, output power, and efficiency. The SMF can enhance the Brillouin gain, leading to higher output power and improved laser performance. It can also contribute to stability by providing feedback and suppressing unwanted cavity fluctuations. Additionally, the SMF may introduce dispersion effects, which can influence the laser's pulse duration, chirp, or spectral width[21].

However, the specific impact of adding an SMF to a Brillouin fiber laser depends on factors like the SMF's properties, laser cavity design, and intended application[22]. Careful consideration and optimization are necessary to achieve the desired performance enhancements or modifications[23].

2. Experimental setup (Methodology)

Figure 1. illustrates the SBS laser oscillator The nonlinear gain is achieved by combining 4.5 meters of erbium-doped fiber (EDF) with 150 meters of single-mode fiber (SMF). The Erbium-doped fiber (EDF) has a linear gain that amplifies the Brillouin signal and lowers the threshold for SBS generation. While increase the non-linearity for the cavity by using the SMF. When the gain exceeds the loss in the ring resonator, the Stokes wave is formed and oscillated to produce a Brillouin fiber laser (BFL) in the reverse direction.

The gain medium (EDF) I-25 (980-125) cut off wavelength (nm) is (900-970) and numerical aperture is (0.23-0.26). Using a WDM Coupler, it was backward pumped by a 980 nm Laser Diode (LD:LC96A74P-20R). It measures 4.5 meters in length and boasts a high concentration of erbium ions. In order to combine the pump from a 980 nm laser diode with the Brillouin pump (BP) signal at 1550 nm, a wavelength division multiplexer, also known as WDM, is utilized. Within the ring resonator, the EDF is utilized to provide assistance for the creation of SBS. In order to tap into the output of the laser, a 20 dB output coupler is utilized. An optical spectrum analyzer (OSA) with a resolution of 0.015 nm is used to characterize the output laser's spectral characteristics, and a digital oscilloscope is used to describe the output laser's temporal characteristics.

Inside the ring cavity, an optical isolator is utilized to prevent the BP from oscillating in the resonator. Because the cavity layout used allows for the counter-propagation of the pump and Stokes fields, Kerr four-wave mixing and cascaded SBS may be avoided. When a highly concentrated EDF is utilized to generate BEFL, a total cavity loss can be reduced, which in turn results in an increase in output power. Additionally, the EDF possesses a significant nonlinearity, which enables the generation of steady pulses at around the Brillouin threshold pump power.

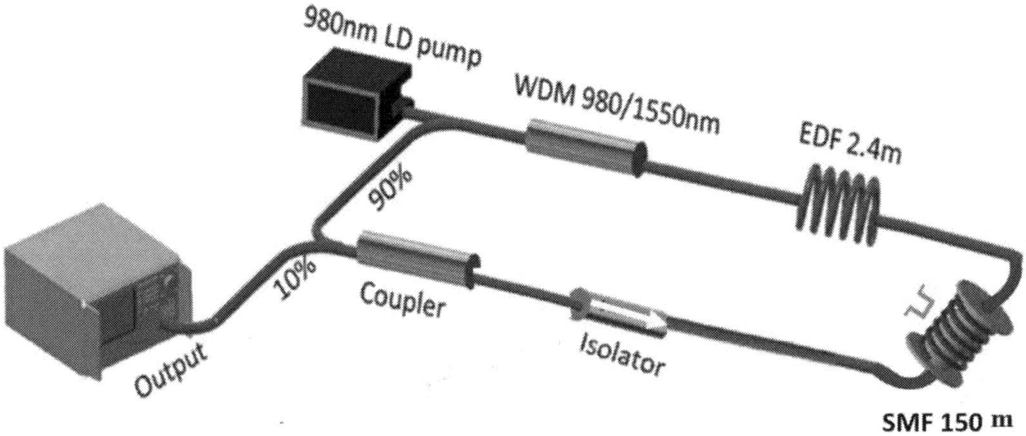

Figure 1. Schematic of the proposed BEFL

3. Results and Discussion

At a pump power of 94 mW, the threshold of Brillouin Fiber Laser (BFL) operating at a wavelength of 1565.916 nm is seen. With the power of the pump increasing from 10 mW to 236 mW, a passive BFL is detected.

The spectral and temporal performances of the Brillouin Fiber laser are depicted in **Figure 2. (a)** Depicts the laser output wavelength of the cavity operation produced with 150 lengths of (SMF) at a maximum pump power of 236 mW as shown in **Figure 2. (b)**. After incorporating the SMF into the cavity, the peak laser wavelength was blue-shifted from 1565.904 nm to 1556.64 nm by 0.08 nm and red-shifted from 1565.9.409 nm by 0.08 nm. A spectral bandwidth of 0.08 nm at -3.15 dB is available from the Q-switched laser. Since the loss increment in the cavity causes the blue- and red-shifted wavelengths, the lasing will emit shorter wavelengths in order to increase amplification gain and offset the additional losses. **Figure 2. (c)** depicts the oscilloscope trace of Q-switched pulses at 236 mW pump power. The oscilloscope trace demonstrates that the Q-switching process is extremely consistent across the time span. The temporal profile of two neighbouring pulses is depicted in **Figure 2. (d)**, where the pulse width and pulse interval between the Q-switched pulses are observed to be 28.6 s and 45.72 s, respectively. The pulse interval of 19.7 microseconds corresponds to a repetition rate of 19.7 kilohertz. In addition, the RF spectrum was analyzed to evaluate the pulsing's stability, as shown in **Figure 2. (e)**. At a fundamental frequency of 19.7 kHz, the RF spectrum of the Q-switched pulse demonstrates exceptional stability, with a signal-to-noise ratio (SNR) of around 64.82 dB.

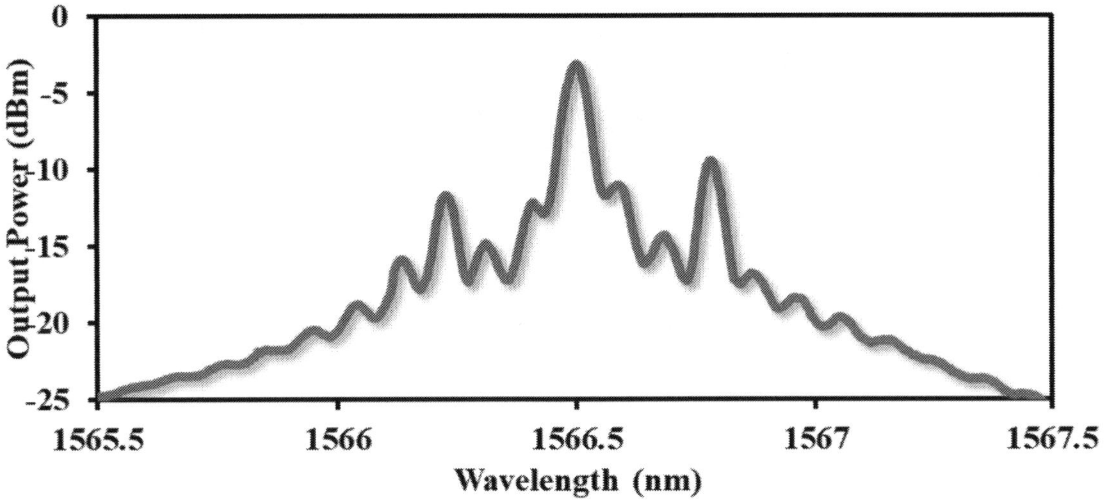

Figure 2. (a) The wavelength of Brillouin Fiber Laser effect.

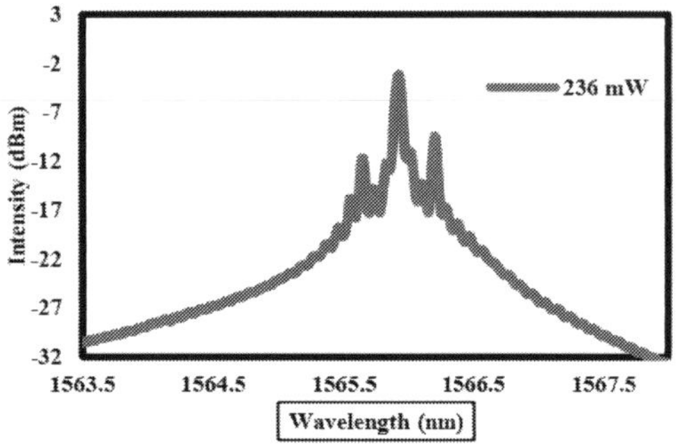

Figure 2. (b) Shown the maximum wavelength at 236 mW pump power

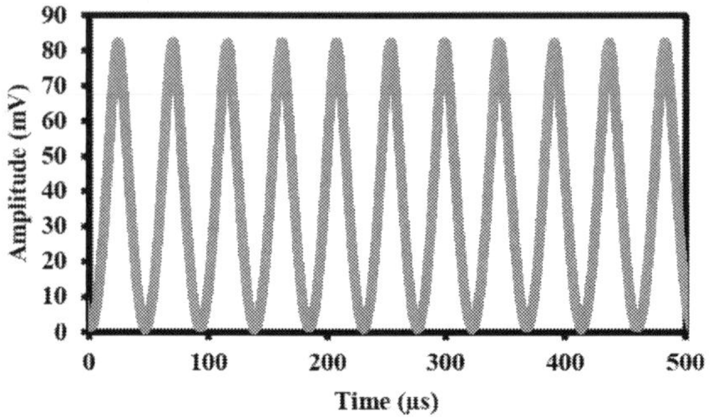

Figure 2. (c) Shown the stable pules of BFL

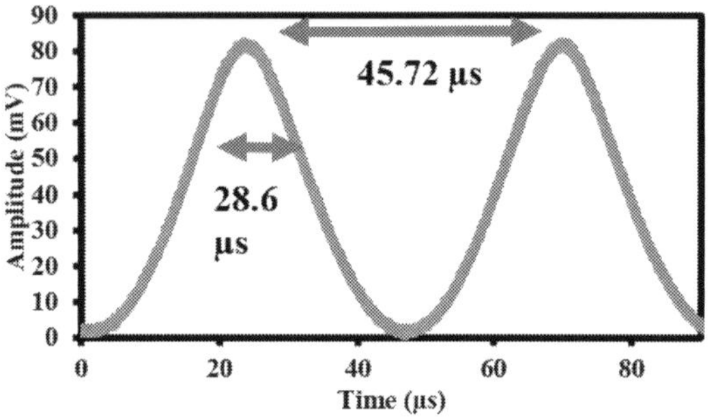

Figure 2. (d) Shown the pulse width and time of and periode for the pulse train

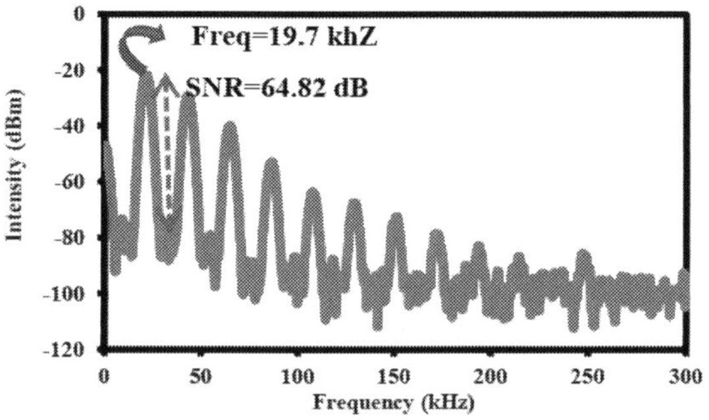

Figure 2. (e) Shown the Radio Frequency the signal noise ratio

During this time, the repetition rates as well as the output power in comparison to the pump power are shown in **Figure 3. (a)**. There was a linearly increasing trend observed in both the repetition rates and the output power as the pump power increased. Finally, the repetition rates climbed from 9.38 kHz to 19.7 kHz, and the output power increased from 1.7 mW to 6.9 mW while the power of the pump grew from 94 mW to 236 mW. Both increases occurred simultaneously. As shown in **Figure 3. (b)** the pulse energy increases from 181.2 nJ to 350.2 nJ along the pump power tuning range, while the pulse width decreases from 64.29 μs to 28.6 μs.

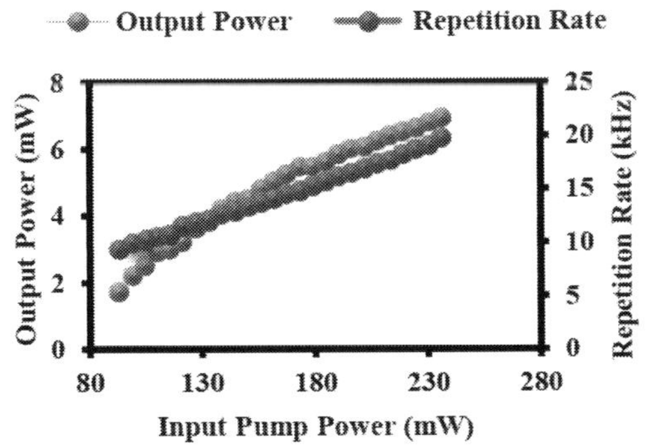

Figure 3. (a) The comparison between the output powerand repetation rate

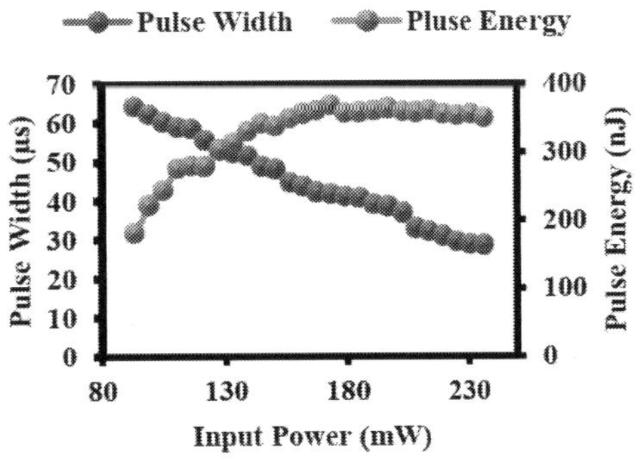

Figure 3. (b) The comparison between the pulse width and pulse energy

4. Conclusion

In this work, we were able to show that a Brillouin fiber laser can be made in SMF by using EDF that has been mechanically exfoliated as the gain medium. It was made in cavity that was set up with a 150 m of SMF. The BFL can be made with the cavity setup.
With the addition of EDF, cavity setup of 150 m of SMF can make BFL work. In this setup, the BFL pulsed better because it had a higher pulse repetition rate and a narrower pulse width. The SMF contained within the ring cavity was utilized to produce self-generating BFL. At a lower pump threshold, the BFL in the linear cavity produced more Brillouin stokes. In the trials, the average spacing was 0.088 nm for the 150 m of SMF at an EDF pump power of 93.468 to 259.493 mW. The entire cavity length of this tiny BFL is 160.34 m. Finally, this is the most recent gain medium length documented for the manufacture of Brillouin erbium fiber lasers without the usage of a Brillouin pump.

Acknowledgments

The research was supported by the Malaysian Ministry of Higher Education (FRGS/1/2021/TK0/UKM/02/17).

References

[1] A. Abass, M. Al-Mansoori, M. Jamaludin, F. Abdullah, T. Al-Mashhadani, and M. Ali, "L-Band multi-wavelength Brillouin–Raman fiber laser with 20-GHz channel spacing," *Fiber and Integrated Optics,* vol. 33, no. 1-2, pp. 56-67, 2014.

[2] M. Al-Mansoori, B. Bouzid, B. M. Ali, M. Abdullah, and M. Mahdi, "Multi-wavelength Brillouin–Erbium fibre laser in a linear cavity," *Optics communications,* vol. 242, no. 1-3, pp. 209-214, 2004.

[3] H. Ali *et al.*, "kHz pulse generation with Brillouin erbium fiber laser," *Laser Physics,* vol. 33, no. 1, p. 015102, 2022.

[4] H. Ali, N. Arsad, N. Zulkipli, A. Rosol, M. Yasin, and S. Harun, "Q-switching pulse generation in multi-wavelength Brillouin Erbium-doped Fiber laser," *Optics Communications,* vol. 534, p. 129300, 2023.

[5] N. Arsad, M. Li, G. Stewart, and W. Johnstone, "Intra-cavity spectroscopy using amplified spontaneous emission in fiber lasers," *Journal of Lightwave Technology,* vol. 29, no. 5, pp. 782-788, 2010.

[6] N. Arsad and G. Stewart, "Stable, tunable, and single-mode operation of an erbium-doped fibre laser system using a saturable absorber for gas spectroscopy applications," in *Fiber Lasers VI: Technology, Systems, and Applications*, 2009, vol. 7195: SPIE, pp. 495-504.

[7] S. W. Harun, S. Shahi, and H. Ahmad, "Brillouin fiber laser with a 49 cm long Bismuth-based erbium-doped fiber," *Laser Physics Letters,* vol. 7, no. 1, p. 60, 2009.

[8] S. W. Harun, M. R. Shirazi, and H. Ahmad, "A new configuration of multi-wavelength Brillouin fiber laser," *Laser Physics Letters,* vol. 5, no. 1, pp. 48-50, 2008.

[9] S. W. Harun, M. R. Shirazi, and H. Ahmad, "Multiple wavelength Brillouin fiber laser from injection of intense signal light," *Laser Physics Letters,* vol. 4, no. 9, p. 678, 2007.

[10] V. L. Iezzi *et al.*, "Temporal characterization of a multi-wavelength Brillouin–erbium fiber laser," *New Journal of Physics,* vol. 18, no. 5, p. 055003, 2016.

[11] L. Li *et al.*, "Absorption line profile recovery based on TDLS and MEMS micro-mirror for photoacoustic gas sensing," *Spectroscopy and Spectral Analysis,* vol. 31, no. 7, pp. 1814-1818, 2011.

[12] Y. Liu, Y. Shang, X. Yi, R. Guo, and Y. Zheng, "Triple Brillouin frequency spacing Brillouin fiber laser sensor for temperature measurement," *Optical Fiber Technology,* vol. 54, p. 102106, 2020.

[13] R. Parvizi, S. W. Harun, N. M. Ali, N. S. Shahabuddin, and H. Ahmad, "Photonic crystal fiber-based multi-wavelength Brillouin fiber laser with dual-pass amplification configuration," *Chin. Opt. Lett.,* vol. 9, no. 2, p. 021403, 2011.

[14] G. S. Rebuli, L. C. Silva, E. Pedruzzi, A. G. Leal-Junior, M. E. Segatto, and C. E. Castellani, "Spectral optimization of Stokes channels for multi-wavelength Brillouin fiber lasers," *Laser Physics,* vol. 32, no. 12, p. 125102, 2022.

[15] S. Shahi, S. W. Harun, and H. Ahmad, "Multi-wavelength Brillouin fiber laser using Brillouin-Rayleigh scatterings in distributed Raman amplifier," *Laser physics letters,* vol. 6, no. 10, p. 737, 2009.

[16] S. Shahi, S. W. Harun, K. Dimyati, and H. Ahmad, "Brillouin fiber laser with significantly reduced gain medium length operating in L-band region," *Progress In Electromagnetics Research Letters,* vol. 8, pp. 143-149, 2009.

[17] M. R. Shirazi, N. S. Shahabuddin, S. N. Aziz, K. Thambiratnam, S. W. Harun, and H. Ahmad, "A linear cavity Brillouin fiber laser with multiple wavelengths output," *Laser Physics Letters,* vol. 5, no. 5, pp. 361-363, 2008.

[18] S. J. Tan, S. W. Harun, N. M. Ali, M. A. Ismail, and H. Ahmad, "A multi-wavelength Brillouin erbium fiber laser with double Brillouin frequency spacing and Q-switching characteristics," *IEEE Journal of Quantum Electronics,* vol. 49, no. 7, pp. 595-598, 2013.

[19] L.-L. Xu, Y.-Y. Wang, L. Jiang, P.-L. Yang, L. Zhang, and S.-X. Dai, "A low-threshold multiwavelength Brillouin fiber laser with double-frequency spacing based on a small-core fiber," *Chinese Physics B,* vol. 30, no. 8, p. 084210, 2021.

[20] Q. Zhao, L. Pei, L. Wu, T. Ning, J. Zheng, and J. Li, "Wide tuning range and high OSNR self-seeded multi-wavelength Brillouin-erbium fiber laser based on a Lyot filter," *Applied Optics,* vol. 57, no. 36, pp. 10474-10479, 2018.

[21] X. Zhou, C. Ge, Y. Wei, G. Yang, M. Bi, and M. Hu, "A tunable multi-wavelength Brillouin–erbium fiber laser with fourfold Brillouin frequency spacing," *Optical and Quantum Electronics,* vol. 49, pp. 1-8, 2017.

[22] Z. Zhou, Q. Yang, Y. Chen, M. Chen, and Z. Zhang, "Tunable Brillouin-Raman multi-wavelength fiber laser with a linear cavity configuration," *Physica Scripta,* vol. 97, no. 9, p. 095502, 2022.

[23] K. Zhu, J. He, K. Chang, D. Xing, Y. Liu, and Z. Wang, "Tunable and narrow linewidth multi-wavelength Brillouin-erbium fiber laser using dual-wavelength pumping," *Optoelectronics Letters,* vol. 18, no. 6, pp. 349-353, 2022.

Journal of Physics: Conference Series **2627** (2023) 012008 doi:10.1088/1742-6596/2627/1/012008

Centralized Monitoring for Huge FTTH Network using Code Grating

***Norhidayah Ahmad[1,2], MS Rahman[1], AH Ahmad Abas[1,2]**

[1] Departments of Electrical and Electronic Engineering, Faculty of Engineering and Built Environment, Universiti Kebangsaan Malaysia (UKM) 43600 Bangi, Selangor.
[2] School of Engineering, Faculty of Computing and Engineering, Quest International University, 30250 Ipoh, Perak.

p106855@siswa.ukm.edu.my

Abstract. The Covid-19 has impacted the demand in broadband communication specifically in Fiber to the Home (FTTH). The increase in the demand caused the increase in the number of users in Triple play service during the Movement Control Order (MCO) period. Consequently, the high volume of user has slowed down the internet speed and connectivity. In addition, there is no effective monitoring system to monitor the network health from user end. This study was conducted to design a PON Fiber-to-the-Home (FTTH) network which can serve thousands of users by increasing the capacity of the network and at the same time the network operator will be able to monitor any faulty branch(s) at the user end. By upgrading the new component in PON FTTH such as SOA, AWG and Demultiplexer, the network is able to support up to 30 000 users and also able to monitor fault user branches in FTTH network by using code FBG.

1. Introduction

When Covid-19 pandemic strike Malaysia in 2020, the network operator experience very high demand in broadband communication service and it been reported that 50% to 70% increase of internet traffic data connected in the residential areas [1]. With drastically increase the number of users in Triple play service during the Movement Control Order (MCO) period has slowed down the internet speed [2]. Due to this, total of 32, 485 complaints received on poor or no 4G connection coverage in July to October 2021 period [3].

Passive optical network (PON) based FTTH access network is the best structure which can provide a higher bandwidth to the users [4,5]. Fiber Bragg Grating (FBG) is low-cost filter which will attached with optical fiber. FBG operation using principle of diffraction grating where when the light travels inside the grating structure some light will reflect back as Bragg wavelength. The wavelength of FBG (Bragg wavelength) is given by equation (1).

$$\lambda_B = 2n_{eff}\Lambda \qquad (1)$$

Where Λ is the grating period, n_{eff} is the effective core refractive index and λ_B is the Bragg wavelength. When Bragg condition is met, all the reflected light constructively adds up to form backward reflected peak whose central wavelength is given by λ_B [6]. FBG as detection and identify

Content from this work may be used under the terms of the Creative Commons Attribution 3.0 licence. Any further distribution of this work must maintain attribution to the author(s) and the title of the work, journal citation and DOI.
Published under licence by IOP Publishing Ltd

failure branches is the most cost-effective monitoring solution and suitable for P2MP access network [7]. The FBG been coded using binary code to represent each ONU [8]. Any faulty ONU can be observed in OSA and analysis by using the code. This method able to monitor up to 256 ONU with different FBG binary code. However, this method needs to used more than one FBG in one branch to represent each ONU. Total up to 512 FBG needed to monitor 256 ONU. A Uniform and Phase-Shifted Fiber Bragg Grating used to monitor the faulty branches [9]. This method used low cost super luminescent diode (SLED) as the monitoring signal. This method able to detect and identified the fault location, however the use of SLED is difficult because the input power is low and it may affect the receiver if the transmitter and receiver is far and can only monitor several users.

The purpose of the study is to design and propose an effective monitoring system for PON H-FTTH network by using coded FBG to address the challenges faced by network operators.

2. Design Consideration

In this paper, a centralized monitoring system with low power monitoring signal and broad bandwidth (Huge Fiber to the Home) for WDM PON is proposed as per shown in Figure 1. The PON FTTH network has been modified by using Optisystem 12 software simulation.

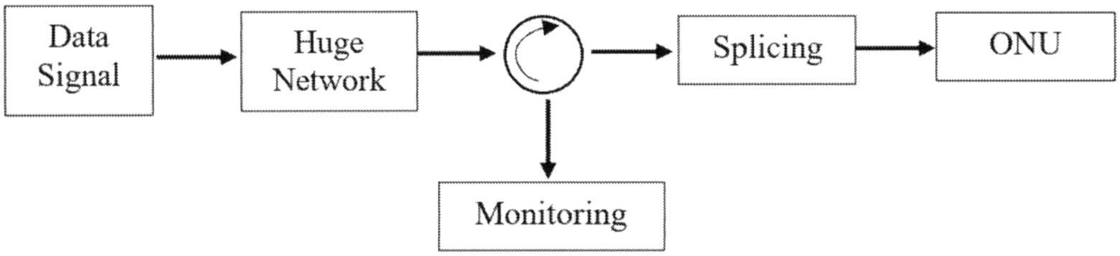

Figure 1. Block diagram for Huge FTTH with monitoring system.

At each ONU will be place one FBG as monitoring system. The reflected wavelength from the FBG is coded by the position of each ONU. The reflected spectrum will be analyzed using OSA. This reflected wavelength will be analysis. If there is no reflected wavelength, its show that there is a faulty ONU and the data was not been transmitted to user.

For DP the FBG will be coded by 1601.0 nm, 1602.0 nm, 1603.0 nm indicate DP 1 until DP 3 respectively. For ONU, each DP will have 8 ONU for example, the number of ONU 1 until ONU 8 will be coded as 1601.1 nm, 1601.2 nm, 1601.3 nm respectively as show in Figure 4. The reflectivity and bandwidth also will be varying depend on the DP and ONU numbers for display purposed in OSA. DP's reflectivity is 0.9 and bandwidth 0.01nm while for ONU the reflectivity is 0.8 and bandwidth 0.02nm as shown in Figure 2. The higher the reflectivity number the better the of reflected signal been reflect at core network. And for bandwidth, the larger number of bandwidths will cause the reflected signal been overlap and it is difficult for the operator to distinguish which ONU is broken.

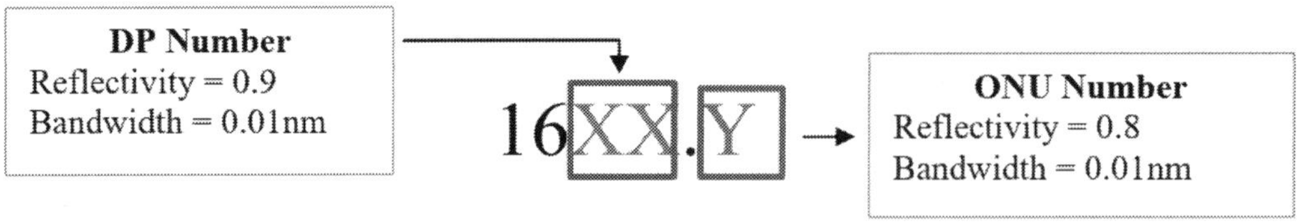

Figure 2. FBG coded

3. Result and Discussion

The parameters investigate in this paper is the number of users, the Q factor, Bit Error Rate (BER) for the user and the reflected FBG signal. The maximum user can the purposed network can support is up to 81 900 users. For purposed of simulation testing, 2 number of DP (DP 1 & DP2) and 4 number of ONU under DP1 was testing. The code FBG wavelength for DP are 1601.0 nm and 1602.0 nm for DP 1 and DP 2 respectively. And code FBG wavelength ONU are 1601.1 nm, 1601.2 nm, 1601.3 nm and 1601.4 nm for ONU1, ONU2, ONU3 and ONU respectively.

Figure 3 show all reflected wavelength from OSA. To verify the system capable to detect faulty branch, attenuator will be placed and act as noise. As show in Figure 4, failure in ONU 3 from DP 1. With this the failure ONU or DP are easily monitor and give advantage to the network operator to do maintenance.

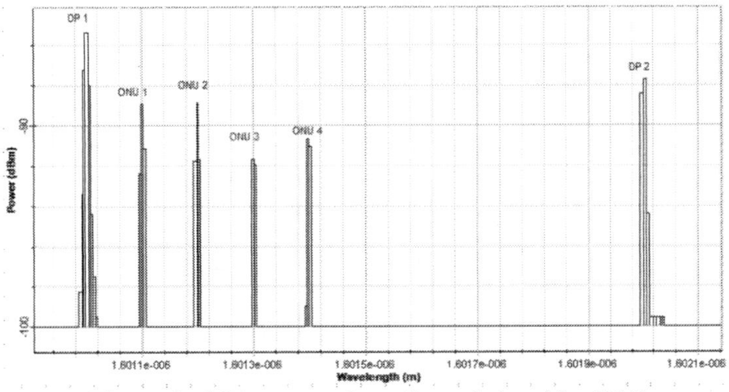

Figure 3. OSA output the reflected signal by FBG

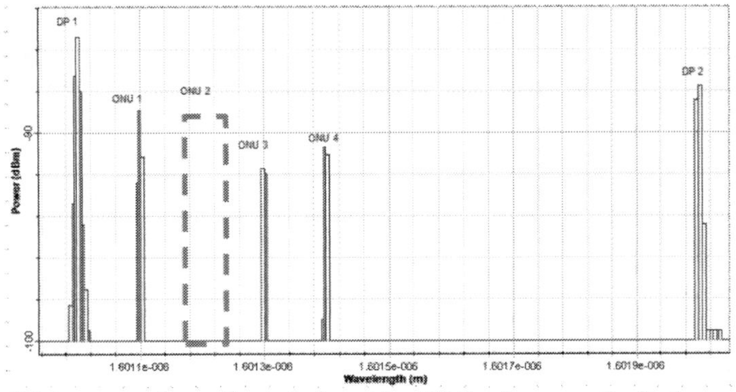

Figure 4. OSA output the reflected signal by FBG when one faulty branch at ONU

The analysis the BER and Q-factor performance the splitting ratio of WDM Demux has been varied to make sure the network can support maximum number of users. With 4 splitting ratio, 10 240 users can use the network with Q-factor 10.2125 and BER performance is 5.58494e-025. While for 32 splitting ratios, Q-factor 9.75384 and BER performance is 5.81494e-023 which can support 81 900 users. From the result, it shows that the higher the splitting ratio, the higher the number of users can support by the system but the lower the Q-factor.

4. Conclusion

A monitoring system for huge FTTH system is successfully presented. In conclusion, the higher the number of users, the lower Q-factor and the smaller the BER value. In increasing the number of users, the monitoring fault(s) branch able to detect by using reflected signal of coded FBG.

5. References

[1] Wong A 2021 *JENDELA: 4G coverage complaints dropped by 34% in Q3 2021, almost 1.5 mil premises fiberised in one year 5:23 pm Comment*
[2] Anon #TECH_ Slow internet speed as more Malaysians go online during MCO
[3] Wong A 2021 *JENDELA: 4G coverage complaints dropped by 34% in Q3 2021, almost 1.5 mil premises fiberised in one year 5:23 pm Comment*
[4] Abdellaoui Z, Dieudonne Y and Aleya A 2021 Design, implementation and evaluation of a Fiber To The Home (FTTH) access network based on a Giga Passive Optical Network GPON *Array* **10** 100058
[5] Al-Quzwini M M 2014 *Design and Implementation of a Fiber to the Home FTTH Access Network based on GPON* vol 92
[6] Sahota J K, Gupta N and Dhawan D 2020 Fiber Bragg grating sensors for monitoring of physical parameters: a comprehensive review *Optical Engineering* **59** 1
[7] Usman A, Zulkifli N, Salim M R, Khairi K and Azmi A I 2020 Optical link monitoring in fibre-to-the-x passive optical network (FTTx PON): A comprehensive survey *Optical Switching and Networking* **39**
[8] Ab-Rahman M S, Ridzuan A M, Kaharudin I H and Hwang I-S *Real Time FTTH Network Monitoring Using Fiber Bragg Grating Codes*
[9] Naim N F, Syuhaimi Ab-Rahman M, Bakarman A H, Ashrif A and Bakar A *Real-time Network Monitoring and Fault Localization in Passive Optical Network using Uniform and Phase-Shifted Fiber Bragg Grating*

The effect of temperature to the reflected signal of PMMA FBG in FTTH network

A H Ahmad Abas[1,2], M S Rahman[1], Norhidayah Ahmad[1,2]

[1] Departments of Electrical and Electronic Engineering, Faculty of Engineering and Built Environment, National University of Malaysia, 43600 Bangi, Selangor
[2] School of Engineering, Faculty Computing and Engineering, Quest International University, 30250 Ipoh, Perak

Abstract. The need for an early detection of disaster has grown vital because of the rise in natural disaster events occurred in Malaysia recently, such as fire, landslides, and floods. Thus, an early warning system could somehow reduce the seriousness of disaster. Due to this, Polymer Optical Fiber (POF) has garnered a lot of interest, because of its promising features especially for becoming a sensing device. This paper study the effectiveness of polymer FBG sensor in Fiber-To-The Home (FTTH) network. The simulation was done using OptiSystem7.0 software. The changes of power(dBm) due to the temperature changes was tabulated. PMMA has shown a significant change in power with averagely 1.123 dBm drop for every 10°C increase of temperature. This indicates that PMMA have promising characteristics to become the main material for fabrication of FBG sensor.

1. Introduction

With recent extreme weather in Malaysia, natural disaster has happened more frequent and unpredicted. Disasters like floods, fires, and landslides have claimed lives while ruining infrastructure and residential buildings. A study has indicate that flood, typhoon, and epidemic disasters significantly reduce the income of the nation [1]. This aftermath of the natural disaster was huge and fatal due to the incompetency of the early detection system.

Due to this, the possibility of using the existing fiber optics network as a sensing device has been raised. The capabilities of FBG sensor in has been studied extensively. It has been studied that the FBG can provide temperature sensitivity of 0.108 nm/ C. [2] [3]. However, the test did not show the changes in power (dBm) of the fiber. By monitoring power changes (dBm) it would be much easier to predict changes in real time. Furthermore, the test was done for a single FBG unit and not in the FTTH network. Thus, this research will simulate the effect of polymer FBG in FTTH network and analyze the effect of temperature change to the reflected signal power (dBm).

1.1. Poly-methyl methacrylate (PMMA)

PMMA is the most often used polymer optical fiber material due to its high quality, transparency, low cost, and ease of processing [4]. In addition, the recycled PMMA can be used for the fabrication of optical fiber [5] and thus reduce the cost of fabrication and promotes sustainability.

In addition to that, the ability to dope PMMA for the development of fiber-based devices such as lasers and amplifiers, as well as the opportunity to improve photosensitivity for effective fiber Bragg

Content from this work may be used under the terms of the Creative Commons Attribution 3.0 licence. Any further distribution of this work must maintain attribution to the author(s) and the title of the work, journal citation and DOI.
Published under licence by IOP Publishing Ltd

grating sensing has made PMMA the best choice in FBG fabrication [6] . The key advantage of adopting PMMA polymer is that it is extremely temperature sensitive [7]. The fundamental downside of PMMA is its high attenuation losses, which can reach 100 dBm in the 1300-1650 nm wavelength interval [8], which is commonly employed in optical communication applications.

2. Simulation design

The design for the sensing system uses FTTH network. The FBG sensor is located at each of the customer-end (ONU). The FBG is being connected in series configurations. The transmitted signals are 1310nm, 1490nm, and 1550nm for triple play service and a white light with 1625nm wavelength is being transmitted for the purpose of monitoring. This simulation uses 20km fiber cables which then ended up at the customer Optical Network Unit (ONU). The change of temperature is being simulated by changing the refractive index of the FBG and the reflected power is monitored.

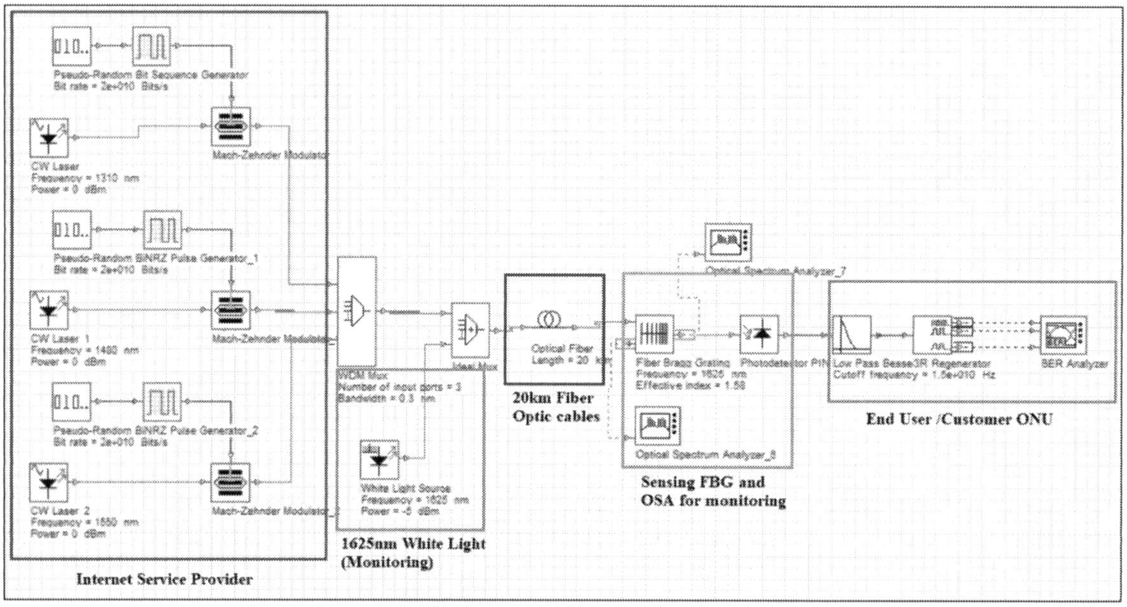

Figure 1. Simulation of POFBG sensing in FTTH network using OptiSystem7.0

2.1. Simulation of PMMA as FBG

In order to simulates PMMA as an FBG material in Optisystem7.0, the refractive index of PMMA for certain temperature values is tabulated based on the study done by M Kovacevic et. Al [3]. This study provides the data of refractive index *vs* temperature changes of PMMA polymer.

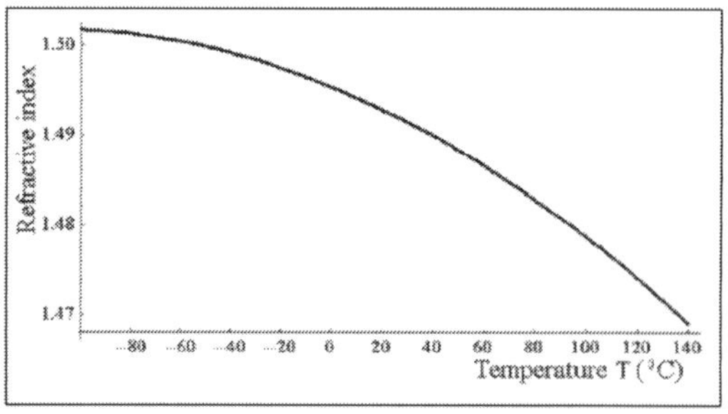

Figure2. The relationship between PMMA refractive index and temperature.

Table 1. The FBG reflected signal power reaction against Temperature changes

Temperature	R.I.	Reflected Power(dBm)	Power Difference(dBm)
20°C	1.492	45.838	
30°C	1.491	44.507	1.331
40°C	1.490	43.523	0.984
50°C	1.488	41.916	1.607
60°C	1.487	41.135	0.781
70°C	1.485	39.692	1.443
80°C	1.483	38.641	1.051
90°C	1.481	37.861	0.780
100°C	1.479	36.849	1.012

3. Results and Discussion

Based on the simulation results of temperature ranges from 20°C to 100°C, the reflected signal power is showing a decreasing trend. This indicates that the higher the temperature, the lower the reflected power of PMMA FBG. In general, when a polymer's temperature rises, its refractive index decreases. This behavior can be explained by the polymer's thermal expansion and the resulting changes in density and molecular interactions. The refractive index values are important because they will affect the reflected power of the FBG, which will then give the indication of temperature changes.

The thermo-optic coefficient can be used to quantify the temperature dependency of the refractive index. The change in refractive index per unit change in temperature is represented by thermo-optic coefficient. Different polymers can have varied thermo-optic coefficient values, and it is frequently stated for a specific temperature range.

The reflected power is monitored by Optical Spectrum Analyzer (OSA). This reflected power can be monitored in real-time, so early prediction of fire (in specific location) can be done by installing PMMA FBG at each ONU location. However, we discovered that the reduction in power loss does not correspond to a drop in temperature. For example, the power reduction from 20°C to 30°C is 1.331dBm, while it is just 0.983dBm from 30°C to 40°C. This shows us that the relationship of refractive index with temperature is not perfectly linear. Nonetheless, this is not a severe issue because a monitoring system should capture any changes and any little difference should be clearly noticeable.

4. Conclusions

PMMA has been investigated for use as a sensing element in FBG. Based on the simulation results, we believe that any decrease in reflected power should provide us with an early warning of a potential fire at a specific location (customer ONU). It is concluded that PMMA exhibits a considerable change in the FBG reflected signal power, with a reduction of 1.123dBm(average) for every 10°C increase in temperature. This is consistent with the study done by Beadie et al.[9] where the PMMA had shown the same behavior. This suggests that PMMA has great potential to become the primary material for the development of FBG sensors. In addition to that, Ren et al.[10] has experimented the possibilities to further enlarge the refractive index changes, which will then increase the sensing sensitivity. Nonetheless, it is important to note that the temperature dependency of the refractive index is not uniform for all polymers and might vary based on factors such as polymer composition and molecular structure.

References

[1] M. I. Qureshi, R. Yusoff, S. S. Hishan, A. S. A. F. Alam, and K. Zaman, "Natural disasters and Malaysian economic growth : policy reforms for disasters management," no. December 1965, 2019.

[2] L. Jin *et al.*, "An Embedded FBG Sensor for Simultaneous Measurement of Stress and Temperature," vol. 18, no. 1, pp. 154–156, 2006.

[3] M. Kovacevic and A. Djordjevich, "Temperature Dependence Analysis of Mode Dispersion in Step-Index Polymer Optical Fibers," vol. 116, no. 4, pp. 649–651, 2009.

[4] B. Zhang and M. Kahrizi, "High-temperature resistance Fiber Bragg grating temperature sensor fabrication," *IEEE Sens. J.*, vol. 7, no. 4, pp. 586–591, 2007, doi: 10.1109/JSEN.2007.891941.

[5] A. D. R. P. Rado *et al.*, "Polymethyl methacrylate (PMMA) recycling for the production of optical fiber sensor systems," vol. 25, no. 24, pp. 71–80, 2017.

[6] M. Atef, R. Swoboda, and H. Zimmermann, "Gigabit Transmission over PMMA Step-Index Plastic Optical Fiber Using an Optical Receiver for Multilevel Communication," pp. 19–21, 2010.

[7] S. N. Kasarova, N. G. Sultanova, and I. D. Nikolov, "Temperature dependence of refractive characteristics of optical plastics," *J. Phys. Conf. Ser.*, vol. 253, no. 1, 2010, doi: 10.1088/1742-6596/253/1/012028.

[8] E. B. Gibelli, J. Kai, E. E. S. Teotonio, M. C. F. C. Felinto, and H. F. Brito, "Journal of Photochemistry and Photobiology A : Chemistry Photoluminescent PMMA polymer films doped with Eu 3 + - NL -diketonate crown ether complex," *"Journal Photochem. Photobiol. A Chem.*, vol. 251, pp. 154–159, 2013, doi: 10.1016/j.jphotochem.2012.10.015.

[9] G. Beadie, M. Brindza, R. A. Flynn, A. Rosenberg, and J. S. Shirk, "Refractive index measurements of poly(methyl methacrylate) (PMMA) from 04–16 μm," *Appl. Opt.*, vol. 54, no. 31, p. F139, 2015, doi: 10.1364/ao.54.00f139.

[10] H. Ren, S. Xu, and S.-T. Wu, "Gradient polymer network liquid crystal with a large refractive index change," *Opt. Express*, vol. 20, no. 24, p. 26464, 2012, doi: 10.1364/oe.20.026464.

Dual-Wavelength Q-switched Fiber Laser Generation by Incorporating 2D Materials in D-SMS Fiber Structure

M. A. Z. Zaini[1*], N. F. F. M. Saaid[1], M. F. M. Rusdi[1], A. A. Latiff[1*], M. F. A. Rahman[2] and A. H. H. Al-Masoodi[3]

[1] Photonics Engineering Research Group (PERG), Fakulti Kejuruteraan Elektronik dan Kejuruteraan Komputer (FKEKK), Universiti Teknikal Malaysia Melaka (UTeM), 76100 Hang Tuah Jaya, Melaka, Malaysia.
[2] Photonics Engineering Research Group (PERG), Fakulti Teknologi Kejuruteraan Elektrik dan Elektronik (FTKEE), Universiti Teknikal Malaysia Melaka (UTeM), 76100 Hang Tuah Jaya, Melaka, Malaysia.
[3] Electronic and Telecommunication Engineering Department, College of Engineering, The American University of Kurdistan, Duhok 42001, Iraq.

*Email: m022220008@student.utem.edu.my and anasabdullatiff@utem.edu.my

Abstract. We demonstrate a dual-wavelength Q-switched fiber laser generation by inserting a combination of SM-MM-SM fiber and a d-shape structure. This combination is known as D-SMS fiber. By side-polishing an SM fiber with 1-dB loss, the d-shape fiber structure is obtained. Then the d-shape SM fiber is spliced to 25 cm long MM and SM fiber to form a D-SMS fiber structure. After inserting D-SMS fiber, a stable dual-wavelength presence at 1562.8 nm and 1563.8 nm (1 nm wavelength spacing). The pulse train generates once we deposited a graphene oxide solution onto the surface of the d-shape structure. This Q-switching operation mode obtained from 45 mW to 111 mW pump power with a maximum pulse energy of 56 nJ. The repetition rate and pulse width have been recorded from 34 kHz to 55 kHz and from 10.22 μs to 6.47 μs, respectively.

Keywords: Q-switched, Mach-Zehnder Interferometer, saturable absorber, graphene, d-shape

1. Introduction

A passively Q-switched fiber lasers can be used for outstanding intentions like remote sensing and laser processing. Passive Q-switched fiber laser is frequently used due to cost-effective and more robust compared to active Q-switched fiber laser [1]. In other hand, Dual-wavelength fiber laser sources generation has gained tremendous interest in optical sensing and microwave generation [2]. Integration between dual-wavelength laser and Q-switching operation laser creating a compact laser source which be able to open wider applications in new laser technology development. Robustness, compact in size and alignment-free are a great choice for all-fiber based laser technology.

The saturable absorber is the main element in the fiber laser cavity. Graphene is under two-dimensional material has excellent photoelectric properties and due to its unique structures, it can attract too much attention [3] in radiation environments including in outer space [4]. The advantages of graphene are cheap in cost, stable, and excellent in performance of optical properties [5]. Due to this factor, the usage of graphene is widely used as a saturable absorber because of low saturable absorption

Content from this work may be used under the terms of the Creative Commons Attribution 3.0 licence. Any further distribution of this work must maintain attribution to the author(s) and the title of the work, journal citation and DOI.
Published under licence by IOP Publishing Ltd

and high modulation depth. Graphene has some disadvantages in terms of weak light-matter interaction and oxidation issues when the material is under high-energy laser irradiation .

In this paper, we demonstrate the integration of SM-MM-SM fiber and d-shape structure in generating dual-wavelength Q-switched erbium-doped fiber laser (EDFL). This fiber integration is called as D-SMS fiber. The graphene oxide solution is coated onto d-shape surface area to induce saturable absorption characteristics in the D-SMS fiber structure. The SM-MM-SM fiber structure induces the Mach-Zehnder interferometer (MZI) effect. Thus, generates stable dual-wavelength in Q-switching operation.

2. Fiber Laser Configuration

Figure 1(a) illustrates the graphene PVA as a passive saturable absorber and the 2.4 m long erbium-doped fiber (EDF) as a gain medium integrated into the fiber laser ring cavity. The EDF (FIBERCORE, I-25) has a core diameter of 9 μm and a numerical aperture (NA) of 0.16, with an absorption coefficient of 27 dB/m at 980 nm. With 980/1550 wavelength division multiplexing (WDM), a single-mode laser diode pump source at 980 nm wavelength is pumped to the EDF. The one-way direction of laser operation in the ring cavity is established using an isolator. To observe the laser performances, about 10% of the laser intensity is tapped out from the laser cavity via a 10-dB optical coupler. The proposed D-SMS fiber structure is integrated into the laser cavity, between the isolator and optical coupler as in the insert image of **Figure 1(a)**. **Figure 1(b)** shows the length of the d-shape fiber that has been polished and the length of the d-shape region is 1.75 mm with 0.08 mm remaining core and cladding. The total laser cavity length is approximately 7 m long. The continuous-wave spectrum of EDFL with and without D-SMS fiber structure is shown in **Figure 1(c)**. Without D-SMS, the peak laser generated at 1564.9 nm with a cavity efficiency of 7%. Once the D-SMS fiber structure is added, the generated laser experienced a blue shift and peak lasing presence at 1562.6 nm. The total shift difference is approximately 2.3 nm, determining additional insertion loss has been induced from the proposed D-SMS fiber structure.

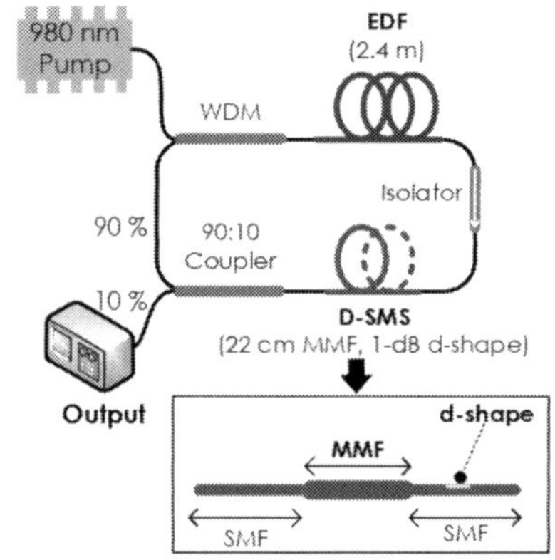

(a)

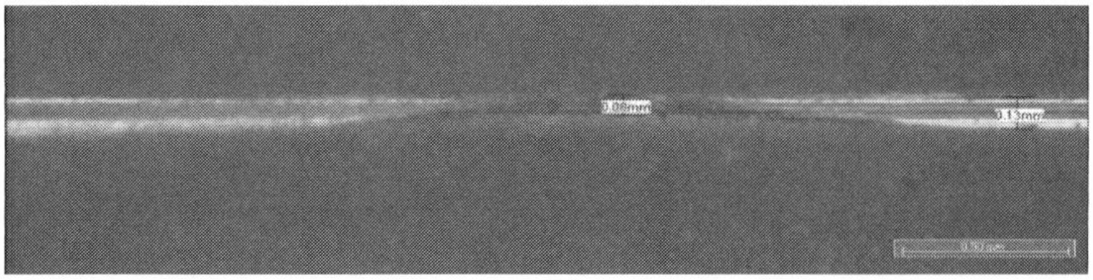

(b)

 2627 (2023) 012010 doi:10.1088/1742-6596/2627/1/012010

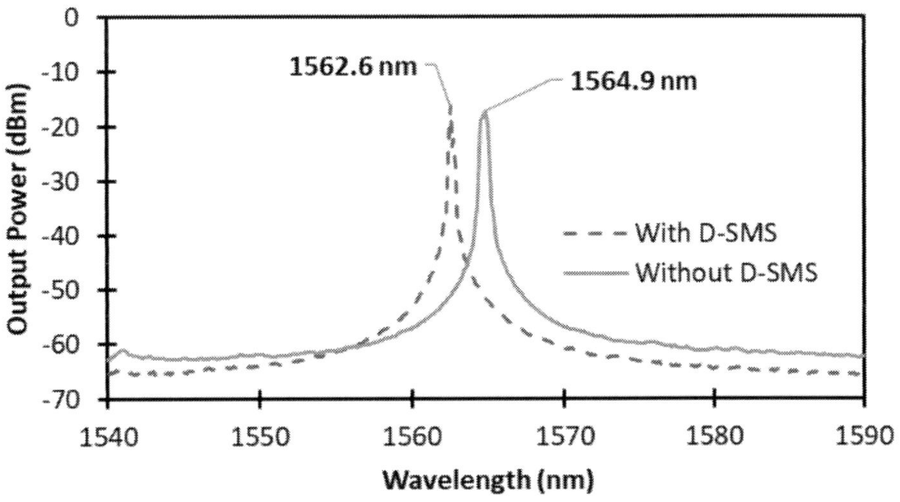

(c)

Figure 1: (a) EDFL ring configuration. Insert image is D-SMS fiber structure. (b) Microscopic image of d-shape fiber. (c) Output spectrum with and without D-SMS fiber structure at threshold pump power.

3. Laser Performance

The laser performance in this paper is reported based on the EDFL with D-SMS fiber structure, where the d-shape surface area is coated with a graphene oxide solution as a passive saturable absorber. This commercially available graphene oxide solution has a 0.05 wt% concentration of graphene oxide and approximately 56% is a carbon element. **Figure 2** shows the measured output spectrum of the dual-wavelength Q-switched EDFL ring cavity. At 52 mW pump power, the lasing was generated at 1562.8 nm and 1563.8 nm. The obtained 1 nm peak wavelength separation is matched with 25 cm long MMF in the D-SMS fiber structure. Thus, meets the MZI condition where the dual-wavelength was observed.

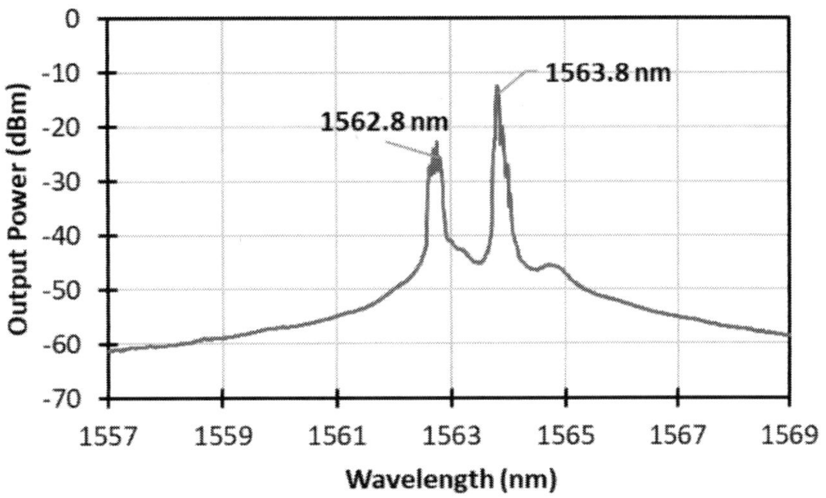

Figure 2: Output spectrum of EDFL with D-SMS fiber structure at 52 mW pump power.

To further characterize the Q-switching operation, the laser output is connected to a 300 MHz Oscilloscope via 1.2 GHz InGaAs Photodetector. **Figure 3** presents the temporal performance analysis of dual-wavelength Q-switched EDFL from 45 mW to 111 mW pump power. The repetition rate and pulsewidth plotting trend in **Figure 3(a)** show the typical trend for Q-switching operation laser (repetition rate: 34 kHz to 55 kHz; pulsewidth: 10.22 µs to 6.47 µs). The shortest pulsewidth is 6.47 µs, obtained at 111 mW pump power. **Figure 3(b)** shows the pulse train of dual-wavelength Q-switched EDFL at 111 mW pump power. As shown in the figure, the measured pulse period is 18.14 µs. Without graphene oxide coated onto the d-shape surface area of the D-SMS fiber structure, no pulse train presence.

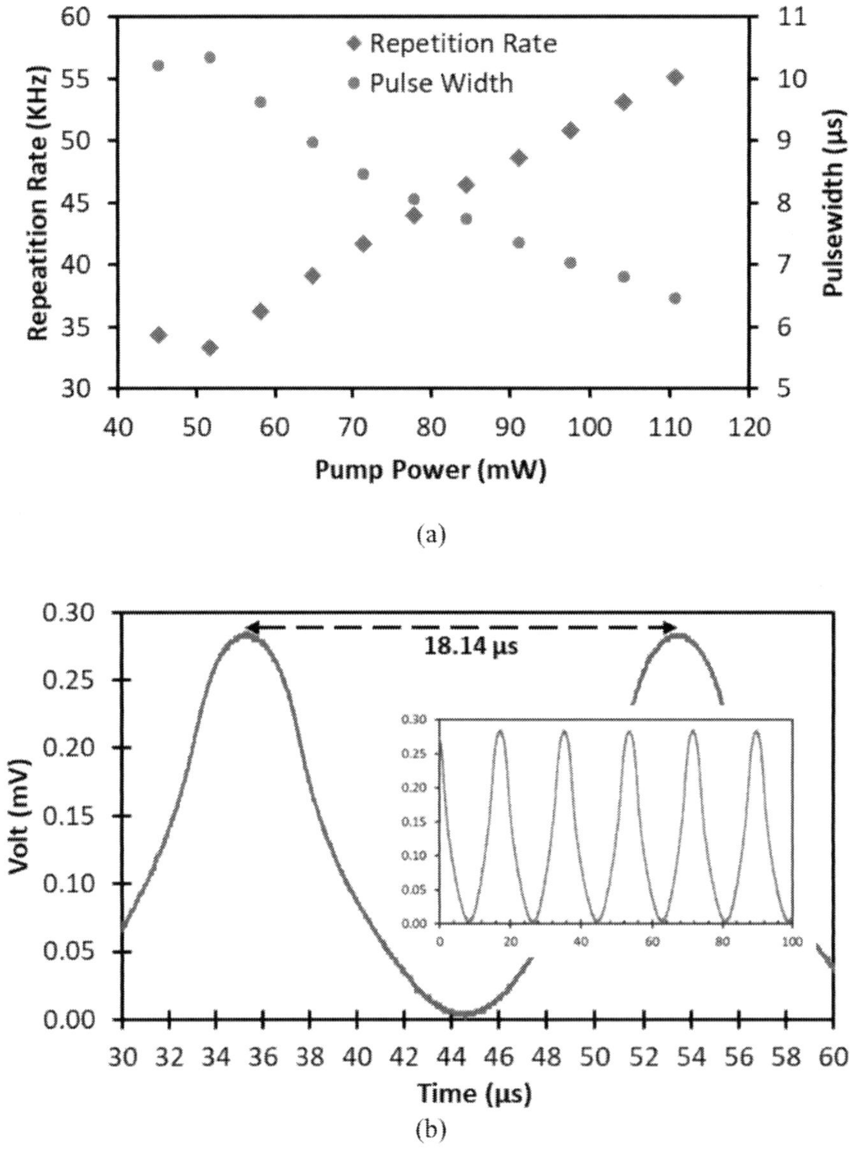

Figure 3: (a) Repetition rate and pulsewidth of EDFL. (b) Pulse train of EDFL at 111 mW.

Figure 4 shows the obtained dual-wavelength Q-switched EDFL has a laser efficiency of 3.4 %. The obtained output power increases from 0.87 mW to 3.11 mW, while the pulse energy relatively is from 25 nJ to 56 nJ. Stable pulsing operation starts at 52 mW pump power. The obtained dual-wavelength Q-switched EDFL has a peak power from 2.45 mW to 8.66 mW. At the maximum pump power, the maximum output power and pulse energy are 3.11 mW and 56 nJ, respectively.

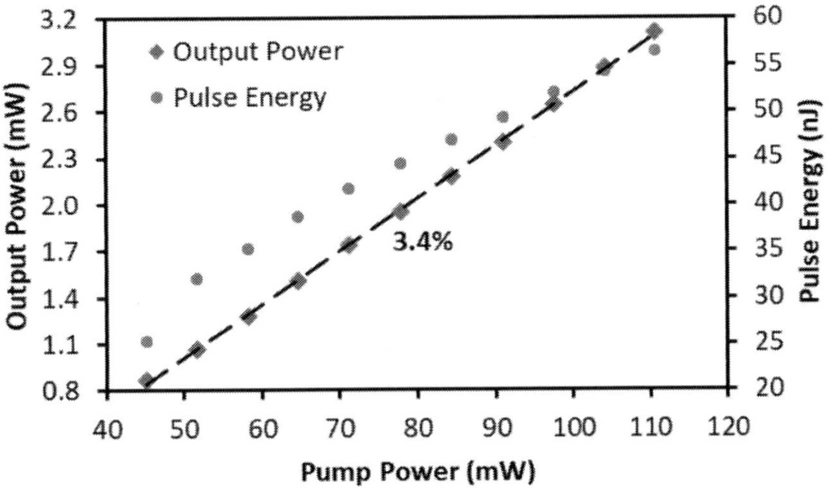

Figure 4: Output power and pulse energy of EDFL.

4. Conclusion

A combination between SM-MM-SM fiber structure and d-shape structure (D-SMS) demonstrates a dual-wavelength EDFL. Once the d-shape surface area of the D-SMS fiber structure is coated with graphene oxide solution, the EDFL laser induces the Q-switching operation from 45 mW to 111 mW pump power. The stable pulsing starts at 52 mW pump power. The D-SMS fiber structure coated with graphene oxide has generated the dual-wavelength Q-switched EDFL with maximum output power, pulse energy, and peak power of 3.11 mW, 56 nJ, and 8.66 mW, respectively.

References

[1] H. Ahmad, H. S. M. Albaqawi, N. Yusoff, W. Y. Chong, and M. Yasin, "Q-Switched Fiber Laser at 1.5 μm Region Using Ti3AlC2 MAX Phase-Based Saturable Absorber," *IEEE J Quantum Electron*, vol. 56, no. 2, p. 1, 2020, doi: 10.1109/JQE.2019.2949798.

[2] A. A. Latiff, N. A. Kadir, E. I. Ismail, H. Shamsuddin, H. Ahmad, and S. W. Harun, "All-fiber dual-wavelength Q-switched and mode-locked EDFL by SMF-THDF-SMF structure as a saturable absorber," *Opt Commun*, vol. 389, pp. 29–34, Apr. 2017, doi: 10.1016/j.optcom.2016.12.011.

[3] S. Liu, H. Huang, J. Lu, N. Xu, J. Qu, and Q. Wen, "Liquid-Phase Exfoliation of Ta2NiS5 and Its Application in Near-Infrared Mode-Locked Fiber Lasers with Evanescent Field Interactions and Passively Q-Switched Bulk Laser," *Nanomaterials*, vol. 12, no. 4, 2022, doi: 10.3390/nano12040695.

[4] D. Kim, N. H. Park, H. Lee, J. Lee, D.-I. Yeom, and J. Kim, "Graphene-based saturable absorber and mode-locked laser behaviors under gamma-ray radiation," *Photonics Res*, vol. 7, no. 7, p. 742, 2019, doi: 10.1364/prj.7.000742.

[5] X. Peng and Y. Yan, “Graphene saturable absorbers applications in fiber lasers,” *Journal of the European Optical Society-Rapid Publications*, vol. 17, no. 1, 2021, doi: 10.1186/s41476-021-00163-w.

Single and dual optical microsphere resonator as Humidity Sensor

Ummu Umairah Mohamad Ali[1,3], Ahmad Ashadi Johari[2], Zulzilawati Jusoh[3], Sulaiman Wadi Harun[4]

[1]Centre for Liberal Studies/Education, Akademi Laut Malaysia, 78200 Melaka, Malaysia
[2]Faculty of Electrical and Electronic Engineering Technology, Universiti Teknikal Malaysia Melaka, 76100 Melaka, Malaysia
[3]School of Electrical Engineering, College of Engineering, University Teknologi Mara Cawangan Terengganu, 21080 Terengganu, Malaysia
[4] Department of Electrical Engineering, Faculty of Engineering,University of Malaya, 50603 Kuala Lumpur, Malaysia

ummu.umairah@alam.edu.my, ashadi@utem.edu.my, zulzilawati@uitm.edu.my, swharun@um.edu.my

Abstract. This study explores into the influence of whispering gallery mode (WGMs) on single and dual microsphere resonator (MSR) as a humidity sensor which can confine light by continuous internal reflection. The construction of the device is simple, and it has a low cost as MSR is made of silica material with a high refractive index. MSR coupling tapered optical fibers with size 5µm to the surfaces of single and two microspheres is demonstrated as a relative humidity (RH) sensor. The Q-factor was measured 7.754×10^5 dBm for single MSR while for dual MSR, Q-factor is 7.765×10^5 dBm. When sensitivity, linearity, and repeatability were analyzed based on transmitted power, the results were conclusively excellent for the performance of the double MSR.

1. Introduction

A whispering gallery mode (WGM) microresonators has been extensively investigated due to their small devices and that can confine the light with high quality factor (Q) [1-3] It is commonly constructed using materials with high refractive index like silica or polymer [4-6]. A WGM microresonators have applications in sensing, laser sources, nonlinear optics, optical filters and switches, and quantum optics [7-10]. They are valuable for their ability to enable precise measurements, compact laser sources, nonlinear optical processes, and exploration of quantum phenomena. These resonators exploit the phenomena of total internal reflection in order to trap and confine light within a circular or spherical cavity, and they are often associated with circular-path resonant cavities such as micro-sphere, micro-ring, and micro-disc geometries. [10-12].

The microsphere resonator's (MSR) ability to confine light in three dimensions using a combination of the

Content from this work may be used under the terms of the Creative Commons Attribution 3.0 licence. Any further distribution of this work must maintain attribution to the author(s) and the title of the work, journal citation and DOI.
Published under licence by IOP Publishing Ltd

WG-bouncing ball and WG ring principles has attracted considerable attention [4, 13]. Two distinct turning points that correspond to regional field enhancement determine the confinement of the WGM in the MSR. The presence of these turning points in MSRs optimises the add/drop operation [14, 15]. MSRs, as opposed to other micro-resonator structures, are capable of generating complex spectra transmitted with highly degenerated resonances [16]. This is accomplished by the overlap of multiple MSR radii, which raises the resonance spectra and traps light near the MSR surface[17, 18].

An optical microsphere humidity sensor is a type of sensor that uses a MSR as a sensing element to measure humidity levels in the surrounding environment. The operation of an optical microsphere humidity sensor is based on the evanescent field sensing principle [10, 15, 19]. When the MSR is exposed to moisture or humidity, molecules of water are absorbed onto its surface. This causes a change in the index of refraction of the MSR, which in turn effects the light transmission through it [15]. To measure humidity, light is coupled into the MSR and allowed to circulate within it using either an optical fiber or waveguide. The light interacts with the water molecules adsorbed on the surface of the microresonator, and the intensity or spectral properties of the transmitted or reflected light are monitored. By analyzing these changes, the humidity level can be determined. These sensors are advantageous such as their great sensitivity and potential miniaturization, making them useful for humidity monitoring in various applications [15]. However, they require careful calibration and environmental control to ensure accurate measurements.

In this work, we have introduced and conducted practical experiments on a fiber sensor that can detect differences in relative humidity. The uses microresonator in humidity sensor utilizes both single and dual microsphere resonators to achieve its functionality.

2. Characterisation of MSR

Standard single-mode fiber (SMF-28) and a fusion splicer machine (Furukawa Electric Fitel S178A) were employed in the manufacture of the microsphere resonators (MSR) for this investigation. A cleaved tip of the SMF-28 was inserted into one of the arms of the fusion splicer so that the MSR could be created. During the process, the equipment that performed the fusion splicing was controlled manually. As a result of the plasma arcs created by the splicer, the fiber tip was heated and partially melted. Because of the surface tension of the molten fiber, it was able to take on a spherical shape, resulting in the formation of a microsphere. It was feasible to adjust the diameter of the microsphere formed by varying the number of plasma arcs delivered to the fiber tip. Figure 1 depicts the MSR structure, which includes a stem with a diameter of 125 µm (Ds) and a microsphere with a diameter of 200 µm (Dm). This MSR is consistent with the dimensions used in previous studies which managed to have the highest in Q-factor compared to other smaller sizes. This manufacturing approach allowed for the precise fabrication of microspheres at the tip of the SMF-28, which supplied the necessary structure for the MSR-based sensing experiments carried out as part of the study [9, 10].

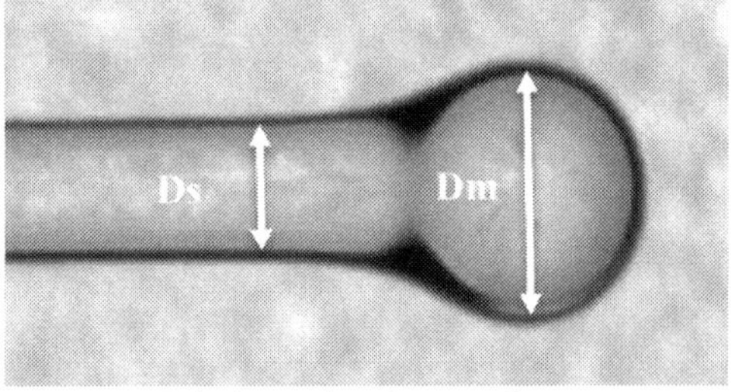

Figure 1. Fabricated micro-sphere resonator

A thinner silica microfiber was manufactured utilizing the flame brushing method to allow for light coupling to and from the microsphere resonator (MSR). The flame brushing method was used to fabricate single-mode fiber (SMF-28) into a non-adiabatic optical microfiber. Thinner microfibers are produced by heating SMF-28 fiber in a flame and pulling it to reduce its diameter. In this study, the manufactured silica microfiber had diameter of 5 µm. Figure 2 (a) show the single MSR coupled to the fabricated microfiber and Figure 2 (b) for the dual MSR. In this configuration, 1550 nm laser light is launched into the microfiber in order to interact with the MSR.

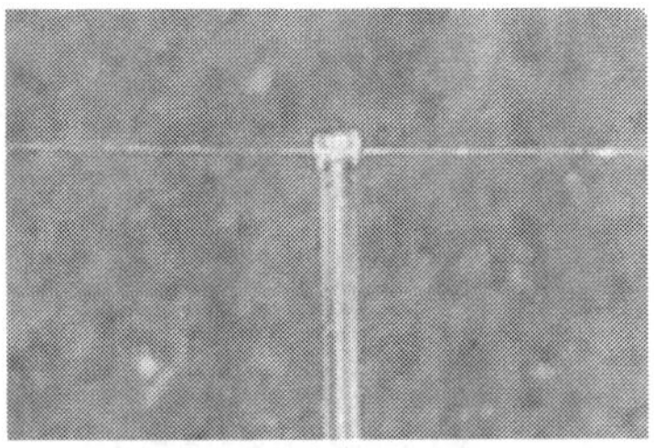

(a)

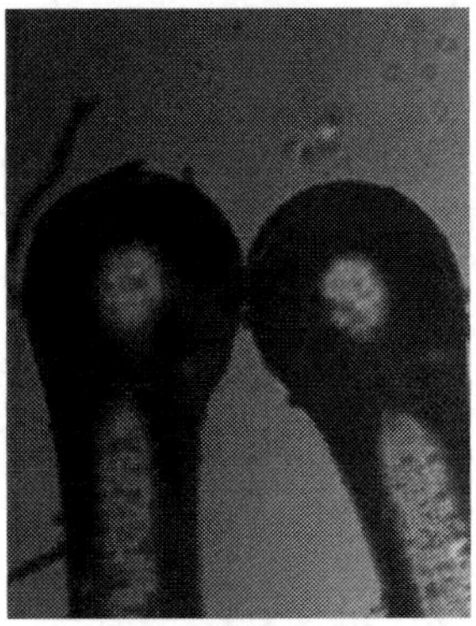

(b)

Figure 2. Tapered microfiber coupled to the a) single MSR b) dual MSR

The single and dual MSR is coupled toward tapered microfiber with a tapering dimension of $5\mu m$ for characterisation procedure. The characterization utilized ANDO AQ4321D tunable laser source (TLS) with wavelength range from 1551.0 nm to 1551.2 nm. The TLS had capability to produce wavelength between 1520 nm until 1620 nm with a 0.001 nm interval was perfectly suited with this experiment's need. The transmitted power wavelength from the tunable laser source will finally be collected using an optical power meter (OPM) from THORLABS S145C was measured as output data. The transmission spectral of single and dual MSR shown in Figure 3 was used to define the quality factor [17, 20]. The Q-factor indicates the quality and effectiveness of the resonant modes within the microspheres by measuring the resonator's capacity to store and sustain light.

Journal of Physics: Conference Series **2627** (2023) 012011 doi:10.1088/1742-6596/2627/1/012011

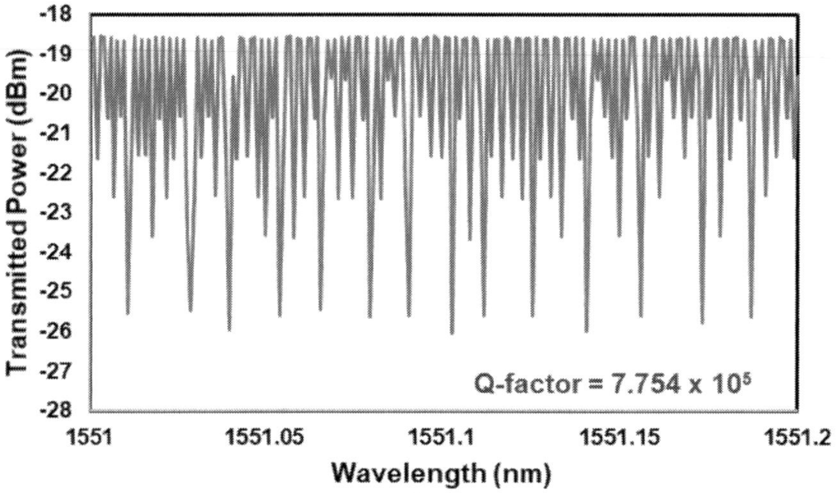

a)

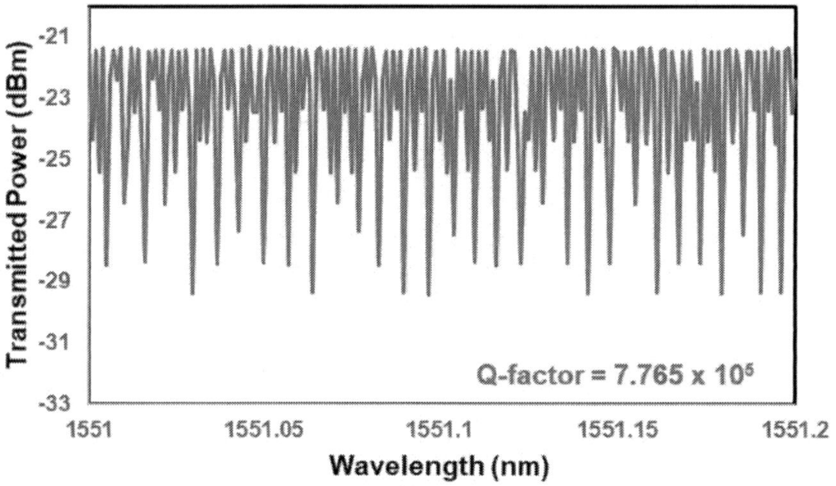

b)

Figure 3. The transmitting spectral of the a) single MSR and b) dual MSR with 5μm tapered microfiber

In the study, the Q-factor measure by formula $Q = \lambda / \Delta\lambda$, whereas $\Delta\lambda$ is the full width half maximum (FWHM) resonant wavelength and λ is represent resonant of the wavelength, is discovered similar compared to the previous work [9, 19, 21, 22] . The Q-factor for single MSR was measured 7.754×10^5 dBm, while Q-factor for dual MSR is 7.765×105 dBm. which higher compared to singe MSR but is slightly not much difference value but still within the same range. The dual MSR had a greater Q-factor than a single MSR may be due to double resonant wavelengths of the two MSRs, causing a slight narrowing in the overall resonance [20]. Additionally, the coupling distance between the resonator and taper microfiber can affect insertion loss. A larger coupling distance can weaken the coupling between the resonator and the tapered microfiber, resulting in a greater insertion loss. On the other hand, a smaller coupling gap can enhance coupling efficacy and decrease insertion loss [23, 24]. The insertion loss for a single MSR experience -19.5 dBm which indicate slightly higher than dual MSR of -21.6 dBm. Due to the coupling gap between the MSR and taper microfiber, the insertion loss may vary between these graph.

3. Performance of Single and dual MSR as Humidity Sensor

The experimental configuration for the MSR used to detect relative humidity as shown in the Figure 4. The key components of the setup include a tunable laser source (TLS), the microsphere resonator (MSR), a relative humidity (RH) measurering meter, and an optical power meter (OPM). The TLS, which serves as a light source for the experiments, is linked to one end of the tapered fibre that has a diameter of 5 μm, while the other end of the tapered fibre is connected to the OPM, which measures the optical power that is transferred through the MSR. The MSR is put inside of a hermetically sealed container in order to create the desired humidity environment. A plate with a salt solution that is completely saturated can be found inside the chamber. This configuration ensures that the MSR is surrounded by an even and consistent level of humidity.

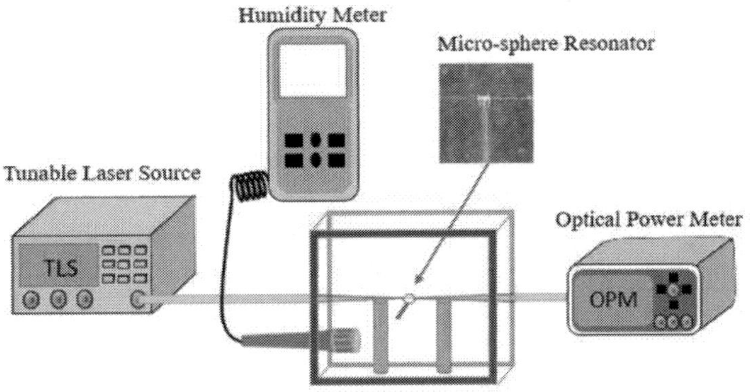

Figure 4. Experimental setup of MSR coupled with tapered microfiber 5 μm

The wavelength was initially set to 1551.103 nm, which was the deepest resonating most resonating depth that was ever recorded for a single MSR. RH is increased from 40% to 80%, and output power is measured at each level. These steps are repeated for dual MSR with a wavelength of 1551.097 which was the deepest resonating most resonating depth that was ever recorded for a dual MSR. In both the single and dual MSR, each RH experiments are repeated three times. All the recorded fiber outputs are depicted in Figure 5, with Figure 5 (a) corresponding to the single MSR and Figure 5 (b) corresponding to the dual MSR configuration. Figure 6 depicts the average of the three test conducted for single and dual MSR.

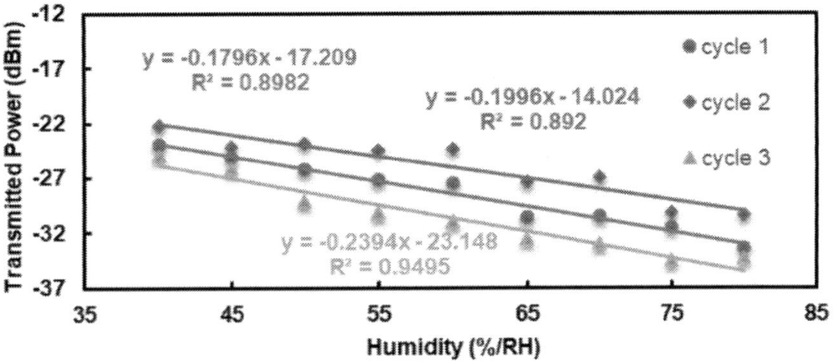

(a)

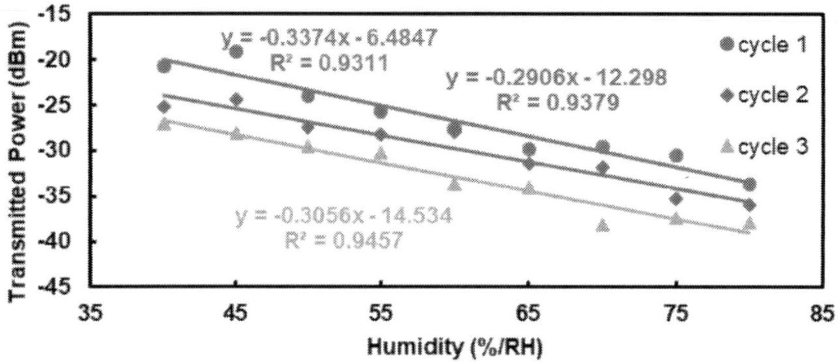

(b)

Figure 5. Repeatability performance of (a) a single MSR (b) dual MSR varies with humidity level

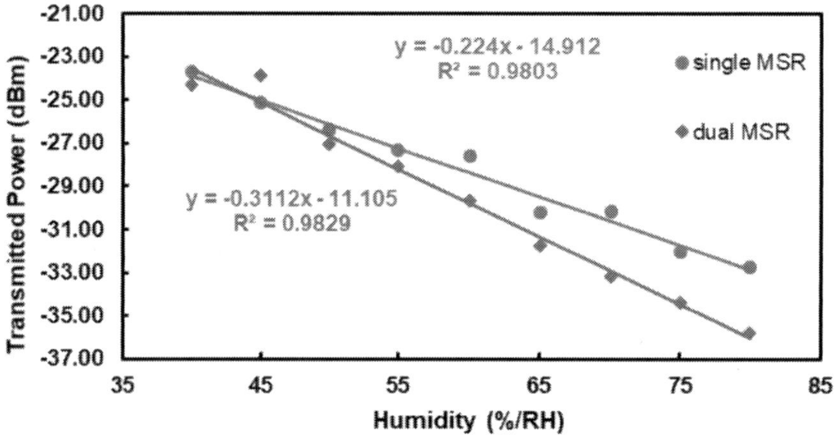

Figure 6. A single and dual MSR coupled with tapered microfiber 5 μm transmission modes

Figure 6 depicts sensitivity and linearity values for single and dual MSR. The sensor's sensitivity describes its capacity to detect and measure minor changes in RH by measuring the corresponding change in output power. While, for linearity is describes the relationship between the output powers of the sensor and various RH percentages. A linear response indicates that the output power is directly proportional to the relative humidity[25]. The sensitivity was measured 0.3112 dB/% RH and linearity at 99.29% for dual MSR, which highest values compare to single MSR. The linearity of the single MSR is 98.03 %, and its sensitivity is 0.2240 dB/%RH. These analyses demonstrated that a dual MSR is superior to a single MBR when employed as a sensor. The performance parameters of the single and dual MSRs are compared and summarized in table 1.

Table 1. Humidity sensing performance for single and dual MSR

Parameters	**Single MSR**	**Dual MSR**
Sensitivity (db/% RH)	0.2240	0.3112
Linearity (%)	98.03	98.29
Linear Range (RH)	40-80	40-80

4. Conclusion

In this research conducted an experimental analysis of both single and multiple MSR implementations as humidity sensors. The resonators were constructed of SMF-28 silica fiber using the method known as "soften-and-compress". The characterisation of the MBRs for the Q-factor value using a microfibre with a diameter of 5μm. Both the single and dual MSR were able to achive a Q-factor of $> 10^5$ which the most significant size contributes the enormous Q-factor value. The single and dual MSR then applied as a humidity sensor where humidity varied from 40% to 80% RH. Via transmitted spectral and wavelength shift, the dual MSR, performed well as humidity sensor. Here, the number of resonators may affect the performance of the resonator as a humidity sensor.

5. References

[1] Y.-L. Chen, W.-L. Jin, Y.-F. Xiao, and X. Zhang, "Measuring the charge of a single dielectric nanoparticle using a high-Q optical microresonator," *Physical Review Applied,* vol. 6, no. 4, p. 044021, 2016.

[2] A. Matsko, A. Savchenkov, D. Strekalov, V. Ilchenko, and L. Maleki, "Review of applications of whispering-gallery mode resonators in photonics and nonlinear optics," *IPN Progress Report,* vol. 42, no. 162, pp. 1-51, 2005.

[3] Y. Zheng *et al.*, "Sensing and lasing applications of whispering gallery mode microresonators," *Opto-Electronic Advances,* vol. 1, no. 9, p. 180015, 2018.

[4] M. Eryürek *et al.*, "Integrated humidity sensor based on SU-8 polymer microdisk microresonator," *Sensors and Actuators B: Chemical,* vol. 242, pp. 1115-1120, 2017.

[5] M. Sumetsky, "Optical bottle microresonators," *Progress in Quantum Electronics,* vol. 64, pp. 1-30, 2019.

[6] M. Cai, O. Painter, and K. J. Vahala, "Observation of critical coupling in a fiber taper to a silica-microsphere whispering-gallery mode system," *Physical review letters,* vol. 85, no. 1, p. 74, 2000.

[7] U. U. M. Ali, A. A. Johari, Z. Jusoh, H. A. Rahman, and S. W. Harun, "A Study of Optical Micro-bottle Resonator Humidity Sensor," in *Journal of Physics: Conference Series*, 2021, vol. 2075, no. 1: IOP Publishing, p. 012017.

[8] N. Pala and M. Shur, "THz photonic and plasmonic devices for sensing and communication applications," in *Micro-and Nanotechnology Sensors, Systems, and Applications XI*, 2019, vol. 10982: SPIE, pp. 478-483.

[9] U. Ali, B. Ahmad, M. Johari, Z. Jusoh, S. W. Harun, and M. Yasin, "Micro-ball and micro-bottle resonators for relative humidity sensing," *Optoelectronics and Advanced Materials, Rapid Communications,* vol. 13, no. 11-12, pp. 567-571, 2019.

[10] M. A. M. Johari, M. I. M. A. Khudus, M. H. B. Jali, A. Al Noman, and S. W. Harun, "Effect of size on single and double optical microbottle resonator humidity sensors," *Sensors and Actuators A: Physical,* vol. 284, pp. 286-291, 2018.

[11] M. Zhang *et al.*, "Broadband electro-optic frequency comb generation in a lithium niobate microring resonator," *Nature,* vol. 568, no. 7752, pp. 373-377, 2019.

[12] A. Pauls, I. Lekavicius, and H. Wang, "Coupling Silicon Vacancies in Diamond to a Silica Microsphere Optical Resonator," in *Laser Science*, 2020: Optica Publishing Group, p. JW6A. 2.

[13] V. S. Ilchenko, M. L. Gorodetsky, X. S. Yao, and L. Maleki, "Microtorus: a high-finesse microcavity with whispering-gallery modes," *Optics Letters,* vol. 26, no. 5, pp. 256-258, 2001.

[14] G. S. Murugan, J. S. Wilkinson, and M. N. Zervas, "Optical excitation and probing of whispering gallery modes in bottle microresonators: potential for all-fiber add–drop filters," *Optics letters,* vol. 35, no. 11, pp. 1893-1895, 2010.

[15] L. Cai, J. Pan, Y. Zhao, J. Wang, and S. Xiao, "Whispering gallery mode optical microresonators: structures and sensing applications," *physica status solidi (a),* vol. 217, no. 6, p. 1900825, 2020.

[16] M. Sumetsky *et al.*, "Surface nanoscale axial photonics: robust fabrication of high-quality-factor microresonators," *Optics letters,* vol. 36, no. 24, pp. 4824-4826, 2011.

[17] M. Pöllinger, D. O'Shea, F. Warken, and A. Rauschenbeutel, "Ultrahigh-Q tunable whispering-gallery-mode microresonator," *Physical review letters,* vol. 103, no. 5, p. 053901, 2009.

[18] H. Huang, *Novel Sensing Mechanisms for Chemical and Bio-sensing Using Whispering Gallery Mode Microresonators*. Washington University in St. Louis, 2018.

[19] U. U. Mohamad Ali *et al.*, "Effect of microfiber diameters on micro-sphere resonator based humidity sensor," *Journal of Electrical and Electronic Systems Research (JEESR),* vol. 20, pp. 66-69, 2022.

[20] M. N. M. Nasir, G. S. Murugan, and M. N. Zervas, "Broadly tunable solid microbottle resonator," in *2016 IEEE Photonics Conference (IPC)*, 2016: IEEE, pp. 759-760.

[21] M. H. Jali *et al.*, "Formaldehyde sensor with enhanced performance using microsphere resonator-coupled ZnO nanorods coated glass," *Optics & Laser Technology,* vol. 139, p. 106853, 2021.

[22] U. Ali, M. Johari, Z. Jusoh, H. Rahman, and S. Harun, "A study of relative humidity sensor on micro-ball resonator," in *Journal of Physics: Conference Series*, 2019, vol. 1371, no. 1: IOP Publishing, p. 012009.

[23] F. Monifi, J. Friedlein, Ş. K. Özdemir, and L. Yang, "A robust and tunable add–drop filter using whispering gallery mode microtoroid resonator," *Journal of Lightwave Technology,* vol. 30, no. 21, pp. 3306-3315, 2012.

[24] M. A. M. Johari *et al.*, "Polyvinyl alcohol coating microbottle resonator on whispering gallery modes for ethanol liquid sensor," *Optics & Laser Technology,* vol. 143, p. 107379, 2021.

[25] C.-Y. Lee and G.-B. Lee, "Humidity sensors: a review," *Sensor Letters,* vol. 3, no. 1-2, pp. 1-15, 2005.

Acknowledgement

This study was supported by RMC Universiti Teknologi Mara (UiTM) Grant (No: 600-RMC/GIP 5/3 (067/2021))

Investigation on the Effect of Fluorescence Material based on D-Shaped Hollow Core Photonic Crystal Fiber

Nur Aina'a Mardhiah Zainuddin[a], Ahmed A. Saleh Falah[a], Muhammad Haniff Mokti[b], Rozalina Zakaria[c], Zulzilawati Jusoh[d], Wei Ru Wong[a]

[a] Integrated Lightwave Research Group, Department of Electrical Engineering, Faculty of Engineering, University of Malaya, Kuala Lumpur 50603, Malaysia

[b] Department of Chemistry, Faculty of Science, Universiti Malaya, 50603 Kuala Lumpur, Malaysia

[c] Photonics Research Centre, University of Malaya, 50603 Kuala Lumpur, Malaysia

[d] Kolej Pengajian Kejuruteraan, Universiti Teknologi MARA, Cawangan Terengganu, Kampus Kuala Terengganu 21080 Kuala Terengganu, Terengganu

Abstract

In this paper, we report the outcomes when fluorescence material coated the hollow core (inner part) of D-shaped photonic crystal fiber (PCF) that has internal diameter (ID), 10 μm. Fluorescence involves the emission of light by a substance that has absorbed light or other electromagnetic radiation. The coated material namely, allene has contributed significantly to the output of transmission spectra at the visible region. The changes in transmission wavelength are remarked when certain parameters vary along the study. The air hole deposited with allene exhibits strong electromagnetic interaction which is good for optical absorption and hence could promote high-sensitivity performance. The optimum condition on how allene covered the hollow PCF which this includes the number of days dipping leads to the bigger wavelength shifting comparing to its emission at 530 nm. Therefore, this study report that allene material could promote an important role in sensing applications at the visible as it displays a good response.

Keywords: fluorescence, hollow core, wavelength shifting.

Introduction

The 1990s witnessed the emergence of photonic crystal fiber (PCF) which later on the application has been developed into sensor purposes (P. S. J. Russell, 2006). The specialty of PCF is the structures (size and geometry of air holes) are easily modified in order for phase-matching conditions. It is proven that the utilization of PCF as a device can improve the sensing performance such as broad detection range (Wang, Lu, Duan, & Yao, 2020) and multi-analyte

Content from this work may be used under the terms of the Creative Commons Attribution 3.0 licence. Any further distribution of this work must maintain attribution to the author(s) and the title of the work, journal citation and DOI.
Published under licence by IOP Publishing Ltd

sensing (Otupiri, Akowuah, & Haxha, 2015) even it has been implemented on gas detection (Bakker & Telting-Diaz, 2002; Moseley, 1997; Shimizu & Egashira, 1999; Smith & Španěl, 2007). Anyhow, the response time of the sensor depends on the capillary action (Kuhlmey, Eggleton, & Wu, 2009) and analyte diffusion mechanisms. Studies report that the D-shaped PCF is more viable as it permits the analyte to penetrate the airholes effortlessly due to its flat surface. Single-mode fiber (SMF) is desirable in this work as a way to eliminate multiple peak losses if other modes are applied (Falah, Wong, & Adikan, 2022). Hollow-core photonic crystal fiber (HC-PCF) comes in two different varieties (Benabid, 2006; Knight, 2003; P. Russell, 2003; P. S. J. Russell, 2006). Here, we proposed a hollow core D-shaped PCF with a single air hole made up of borosilicate. As light travels faster in air compared to the other mediums like glass, hollow-core fiber provides higher speed and lower latency.

Fluorescent molecules absorb photons of specific wavelengths and emit a longer wavelength photon within nanoseconds (Benito-Pena, Valdés, Glahn-Martinez, & Moreno-Bondi, 2016). Recently, fluorescent materials have been widely used in the life and material sciences. Heterocyclic compounds that carry out fluorescent characterisation are useful in bioanalytical applications, including light-emitting diodes (Ullah et al., 2022). Modern heterocyclic molecules are at the core of this development of fluorescence (Schramm & Weiss, 2019). The aromaticity of a heterocyclic compound is what defines a heteroaromatic chemical. Currently, the newly synthesized heteroaromatic compound contains a function group of allenes. Allenes are organic compounds that contain one carbon-centered connected to two adjacent carbons through double bonds. The parent compound of this class is propadiene, which is itself also called allene, also known as the simplest form of allenes.

$$H_2C = C = CH_2$$

where H is a hydrogen and C is a carbon. According to Taylor (1967), the most stable bonding arrangement involves two mutually perpendicular pie bonds, with the central carbon atom (sp-hybridized) joined in a straight line to the two terminal carbon atoms (sp2-hybridized). Today, many patents address the use of allenes as colours, medicines, antioxidants, fibres, and beat- or corrosion-resistant materials, indicating that their industrial potential is being explored.

Methodology

The first part of the project is to prepare the allene material through process of chemistry. As the material is ready, the concentration of allene solution is varied to obtain the emission at 530 nm. Specific concentration at 33 g/L of allene solution is proven to be the best indication has proceeded towards fiber testing. Dipping process take place to allow the diffusion based on the capillary action of the airhole.

A 3.0 cm length hollow core PCF is placed between the light source and optical spectrum analyzer (OSA) as shown in Figure 1 (a). The tested sample with inner coated of allene material by dipping technique has been measured at a specified time. After a proper alignment is done (Figure 1b), the transmission signal is observed at the OSA device while the input to the PCF fiber is a supercontinuum source at wavelength range 400 nm – 2400 nm with ~100mW power. Figure 1 (c) indicate the feature of allene solution in yellow colour.

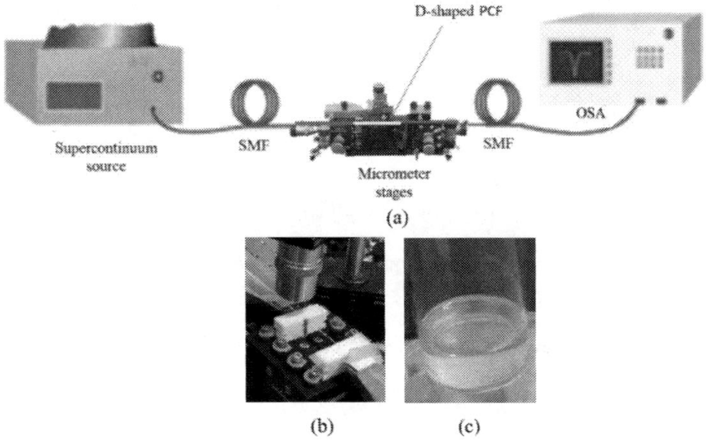

Figure 1. (a) Experimental setup (b) Fiber alignment (c) Allene solution

Results and discussion

At first, the study is about the emission spectra of the material itself which it is called allene. In order to understand the interaction of the material with the light, the optimum concentration of allene has to figure out. Several attempts (Figure 2 (a) and Figure 2 (b)) had been carried out to find the best single peak emission of allene such in Figure 2 (c). Lower concentration producing double peak probably due to not strong or enough emission where the peak arises at wavelength around 450 nm.

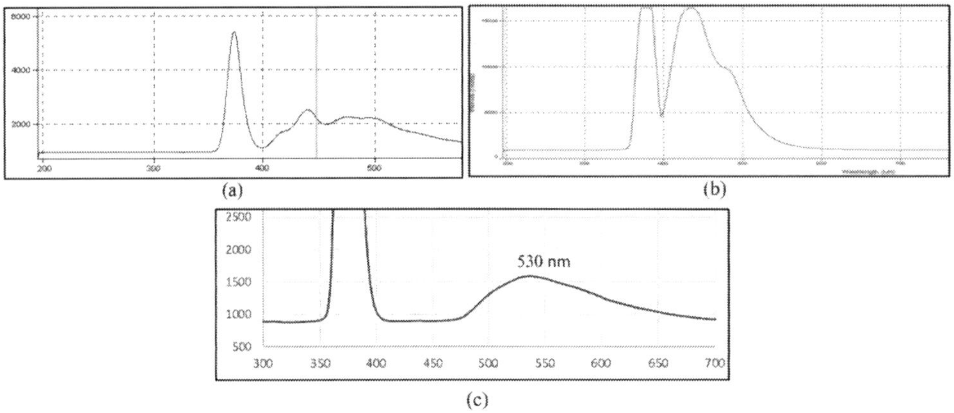

Figure 2: Allene concentration (a) 6 g/L (b) 10 g/L (c) 33 g/L

The outcomes were based on the transmission spectra that showing notable values when the day of dipping fiber in the allene solution are varied. Figure 3 (a) depicts the transmittance result regarding bare fiber, 1-day dip, and 4 days dip. There is a significant shift in wavelength when the days of dipping are increased at 10 nm difference comparing of 1-day dip, 538 nm with 4 days dip, 548 nm. Figure 3 (b) shows the prominent transmittance spectra of 642 nm and 658 nm respectively when second round dipping take place using the previous samples (in Figure 3 (a)). The day of dipping fiber also has been manipulated which show fiber with a minimum day dipping (1-day) has better results in contrast of second time dipping for three days (Figure 3 (c)) that showing no significant spectra probably due to the losses of propagated light throughout the core. Second dipping for about one day exhibits larger shifting, ~ 10 nm from the result in Figure 3 (a). The amount of solution formed inside the hollow core in this period is barely enough for the light to interact well with the surrounding. The other thing that plays a role is the length of fiber testing. The presence of air bubbles (Figure 3 (d)) and the distribution of the solution in the inner fiber approximately at 10.54 μm hollow core (Figure 3 (e)) could affect the transmission.

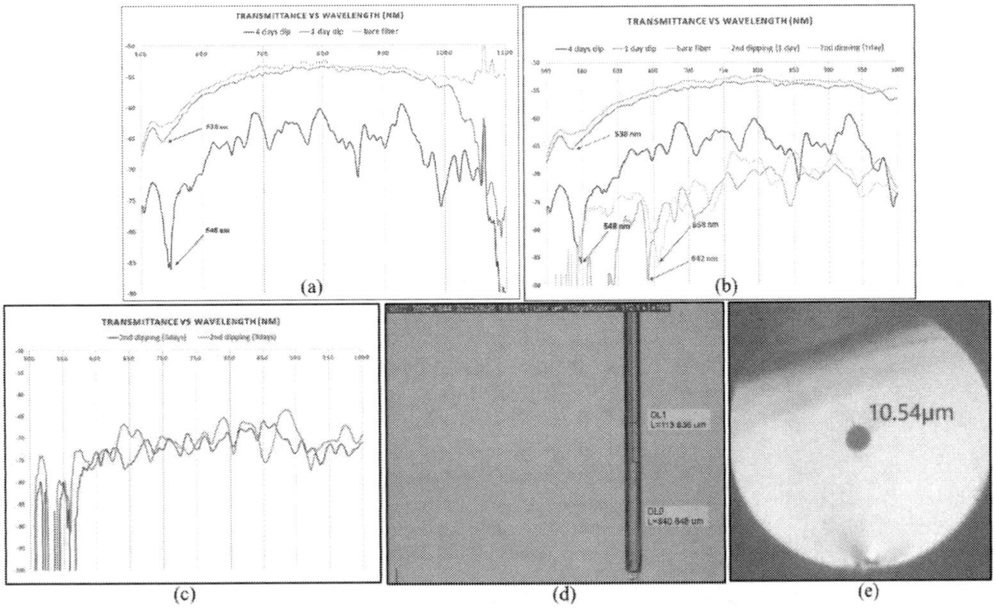

Figure 3: Transmission spectra with number of day dipping (a) first dipping (b) second dipping, 1 day (c) second dipping, 3 days (d) allene distribution inside core (e) ID of hollow core

Conclusion

Experimental work has been carried out using fluorescence material namely allene coated internally of hollow-core photonic crystal fiber (PCF). The optimized condition of the coating area has increased the shifting of transmission spectra. Thus, this study should be well pursued as it is manifest a strong response that could help in sensor applications.

References

Bakker, E., & Telting-Diaz, M. (2002). Electrochemical sensors. Analytical chemistry, 74(12), 2781-2800.

Benabid, F. (2006). Hollow-core photonic bandgap fibre: new light guidance for new science and technology. Philosophical Transactions of the Royal Society A: Mathematical, Physical and Engineering Sciences, 364(1849), 3439-3462.

Benito-Pena, E., Valdés, M. G., Glahn-Martinez, B., & Moreno-Bondi, M. C. (2016). Fluorescence based fiber optic and planar waveguide biosensors. A review. Analytica chimica acta, 943, 17-40.

Falah, A. A. S., Wong, W. R., & Adikan, F. R. M. (2022). Single-mode eccentric-core D-shaped photonic crystal fiber surface plasmon resonance sensor. Optics & Laser Technology, 145, 107474.

Knight, J. C. (2003). Photonic crystal fibres. nature, 424(6950), 847-851.

Kuhlmey, B. T., Eggleton, B. J., & Wu, D. K. (2009). Fluid-filled solid-core photonic bandgap fibers. Journal of Lightwave Technology, 27(11), 1617-1630.

Moseley, P. (1997). Solid state gas sensors. Measurement Science and technology, 8(3), 223.

Otupiri, R., Akowuah, E. K., & Haxha, S. (2015). Multi-channel SPR biosensor based on PCF for multi-analyte sensing applications. Optics Express, 23(12), 15716-15727.

Russell, P. (2003). Photonic crystal fibers. Science, 299(5605), 358-362.

Russell, P. S. J. (2006). Photonic-crystal fibers. Journal of Lightwave Technology, 24(12), 4729-4749.

Schramm, S., & Weiss, D. (2019). Fluorescent heterocycles: Recent trends and new developments. Advances in Heterocyclic Chemistry, 128, 103-179.

Shimizu, Y., & Egashira, M. (1999). Basic aspects and challenges of semiconductor gas sensors. Mrs Bulletin, 24(6), 18-24.

Smith, D., & Španěl, P. (2007). The challenge of breath analysis for clinical diagnosis and therapeutic monitoring. Analyst, 132(5), 390-396.

Ullah, F., Ullah, S., Khan, M. F. A., Mustaqeem, M., Paracha, R. N., Rehman, M. F. u., . . . Bungau, S. (2022). Fluorescent and Phosphorescent Nitrogen-Containing Heterocycles and Crown Ethers: Biological and Pharmaceutical Applications. Molecules, 27(19), 6631.

Wang, G., Lu, Y., Duan, L., & Yao, J. (2020). A refractive index sensor based on PCF with ultra-wide detection range. IEEE Journal of Selected Topics in Quantum Electronics, 27(4), 1-8.

Exploration of Fibre Bragg Grating for Potential Breathing Behaviour Monitoring System

N. A. S. Noorbahrin[1,2], S. N. F. Mohd Asseri[1,2], H. Ahmad[3], K. S. Lim[3], I. Bahari[1,2], K. Thambiratnam[1,2] and M. Z. Zulkifli [1,2,*]

[1]IIUM Photonics and Quantum Research Centre (iPQC), Kulliyyah of Science, International Islamic University of Malaysia, Bandar Indera Mahkota, 25200 Kuantan, Pahang, Malaysia
[2]Physics Department, Kulliyyah of Science, International Islamic University of Malaysia, Bandar Indera Mahkota, 25200 Kuantan, Pahang, Malaysia
[3]Photonics Research Center, University of Malaya, 50603 Kuala Lumpur, Malaysia

Corresponding author: mzz@iium.edu.my

Abstract. An optical-fiber based temperature sensor for the monitoring of breathing behaviour during different physical activities is proposed and demonstrated. The proposed sensor utilizes Fiber Bragg Gratings (FBGs) to detect temperature changes in breath, which are then correlated to the breathing behaviour and thus able to profile the patient for breathing abnormalities. The proposed sensor is cost-effective and light-weight, and can be integrated into wearable devices for real-time monitoring of the patient.

Keywords: Breathing sensor, optical sensor, FBG, temperature

1. Introduction

In monitoring human health, the immediate condition of the human body can be obtained in the form of a snapshot of its vital signs [1]. These vital signs include the body temperature, blood pressure, pulse rate and respiration rate, and any variations or sudden changes in these vital signs can be taken as an early warning for abnormal physical conditions. Respiratory parameters in particular can provide an indication of health abnormalities via an abnormal respiratory assessment [1]. For example, a breathing frequency higher than 27 breaths per minute is an early indication of systemic heart failure, while abnormal respiration can also indicate medical conditions such as pneumothorax or chronic obstructive pulmonary disease [2,3,4]. Typically, health monitoring must be done at medical institutions such as hospitals or clinics due to the need for various sophisticated equipment. However, developments in technology have now enabled home monitoring to be carried out, through the development and commercialization of medical technologies like wearable technologies and smart textiles.

Fiber Optic Sensors (FOS), such as Fiber Bragg Gratings (FBGs), have significant potential to serve as a light, compact and highly accurate respiratory monitoring system. Earlier research has already shown the potential for FBGs as a mechanism to track vital signs via integration into smart clothing in the form of wearables [1,5,6]. This is due to the advantages that the FBG possesses, such as immunity to electromagnetic interference and polarization, being light weight and low cost, as well as its ability to multiplex multiple signals along a single transmission fiber [7]. Silva et al. (2011) designed and fabricated a wearable system based on a single FBG sensor to monitor respiratory rate and health rate, while Tavares et al. [1] demonstrated an FBG monitoring system by embedding the single, bare FBG

Content from this work may be used under the terms of the Creative Commons Attribution 3.0 licence. Any further distribution of this work must maintain attribution to the author(s) and the title of the work, journal citation and DOI.
Published under licence by IOP Publishing Ltd

sensor into the 3D printed material called the Flexible. However, there have only been a few reports of FBG based FOS being used as a respiratory monitoring system.

In this work, a bare FBG sensor was embedded into a respiratory mask to monitor the respiratory behaviour of a test subject after undergoing three different physical activities: resting, walking, and stair climbing. The bare FBG sensor is configured to precisely monitor the temperature of the test subject's breath from the shifted wavelength. The data and results recorded from the FBG temperature sensor were able to distinguish the different between three mentioned physical activities of the test user, thus indicating the potential of the FBG based respirator for developing a breathing sensor.

2. Experimental Setup

2.1. Respiratory Mask and Detection System Design

The breathing behaviour monitoring system was built inside a conventional 12-cm-sized respiratory mask. An FBG calibrated for temperature measurement was used as the FOS. The FBG has an approximate center wavelength of 1544 nm and 99% reflectivity as well as a 3 dB bandwidth of approximately 0.3 nm. All parameters are as obtained at room temperature. The FBG was installed inside the respiratory mask, slightly below the metal piece using clear tape as shown in the red box in Figure 1. A conventional temperature sensor using a J-type thermocouple with a detection range of -210°C to 760°C at 75% accuracy was used as the control sensor.

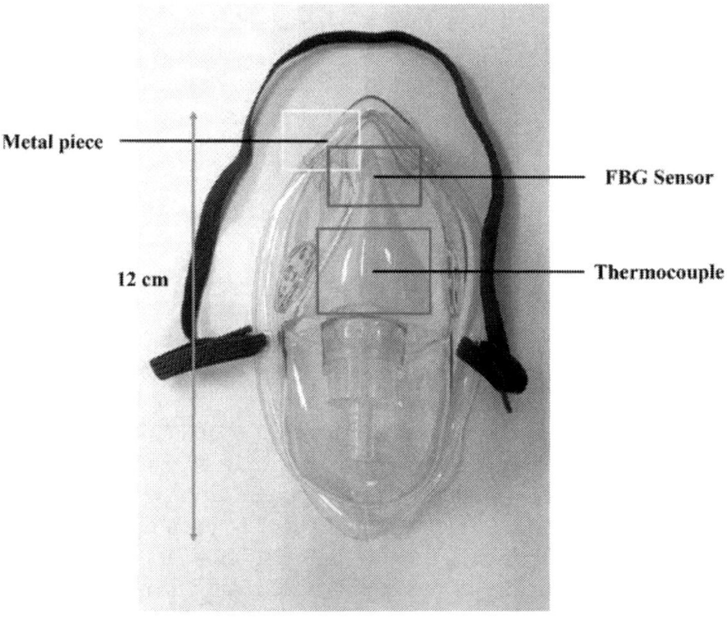

Figure 1. The respiratory mask with FBG labeling of FBG sensor in respiratory mask

The broadband sensor signal was obtained from a C-band erbium-doped fiber (EDF) amplified spontaneous emission (ASE) source. The broadband light source was connected to port 1 of the optical circulator (OC), while the respirator mask with integrated FBG was connected to port 2. Port 3 of the OC was connected to the IBSEN PHOTONICS 1525–1570 nm Interrogator Monitor (I-MON USB) to analyze and record the Bragg wavelength shift from the FBG sensor against time. The Bragg wavelength response approximates the respiration waveform from the inhalation and exhalation air of the test subject. Readings are taken every second up to periods of 15 minutes.

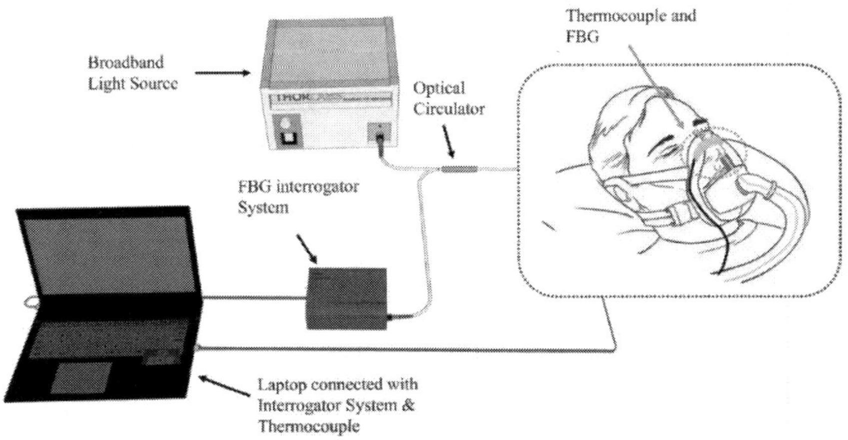

Figure 2. Experimental Setup of breathing behavior using FBG temperature sensor

2.2. *Test On User*

The preliminary assessment of the breathing behaviour monitoring system is evaluated on a test subject with the physical measurements as given in Table 1. The test subject is in good health and is not suffering from any conditions that could cause breathing problems, such as allergies, asthma, or anxiety. In this experiment, the test subject engaged in three different activities: resting, walking, and stair climbing. Each activity was done for 5 minutes before the test subject wore the FBG respiratory mask for respiratory monitoring.

Table 1: Anthropometric characteristics and activities of a test subject

Gender	Male
Age	23
Height (cm)	166
Weight (kg)	55
Activities	Resting, walking, stairs climbing

3. RESULTS AND DISCUSSION

Figure 3 shows the Bragg wavelength shift against the temperature. It can be seen clearly that as the temperature increases from 25oC to 250oC, at 20oC of increments, the Bragg wavelength shifts from 1543.70 nm to 1546.60 nm. The sensitivity of the bare FBG temperature sensor is calculated to be 0.0124 nm/°C, which matches that of similar systems [3,8].

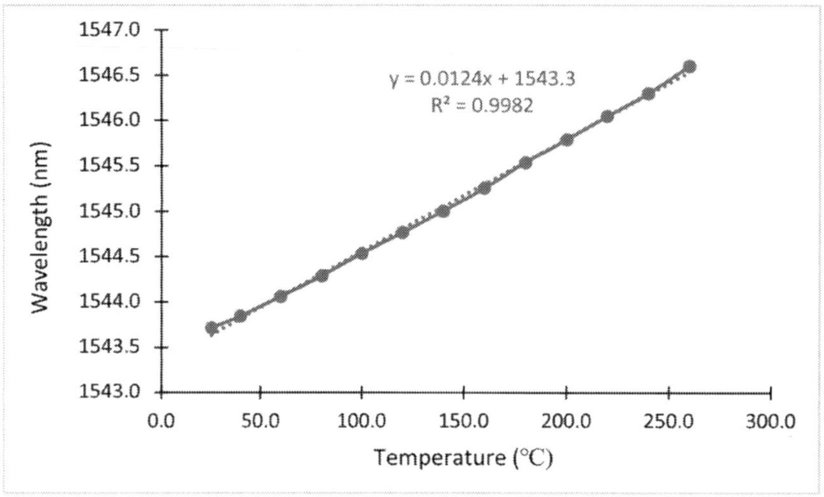

Figure 3: The Bragg wavelength shift against temperature

Figure 4 shows the temperature response taken at intervals of 1 second from both the FBG and thermocouple for each activity: resting, walking, and stairs climbing. It can be seen that the FBG sensor is capable of providing a more accurate reading of the temperature as compared to the thermocouple. From Figure 4(a), the thermocouple provides temperature readings of between 28.3°C and 29.65°C only, while the FBG sensor returns a higher range of detected temperatures from 28°C to 30.5°C. Similarly, from Figure 4(b), the thermocouple detection range is only between 30°C and 30.21°C while the detection range of the FBG is larger, from 30°C to 32.05°C. The readings of Figure 4(c), also provide a similar response, with the FBG having a larger temperature detection range of 29.93°C to 33.55°C as compared to that of the thermocouple which ranges from 29.5°C to 30.2 °C. The possible difference in the detection range could be due to the response time of the FBG as compared to the thermocouple.

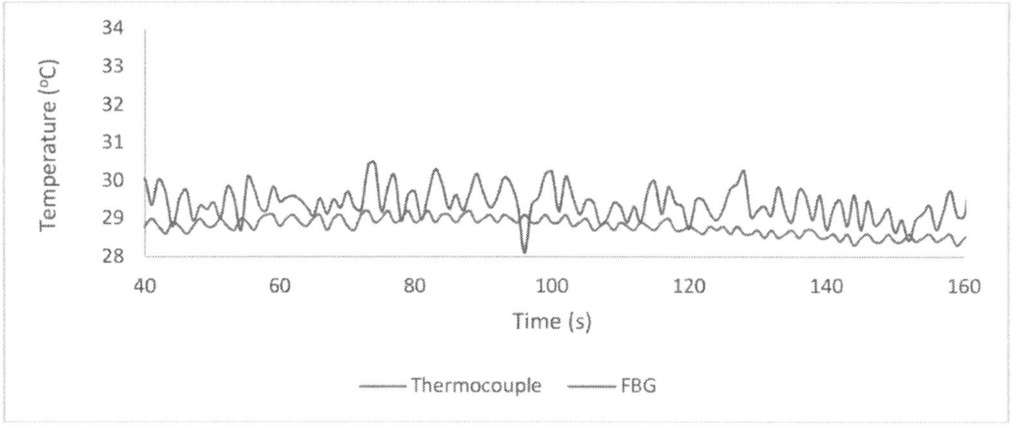

(a)

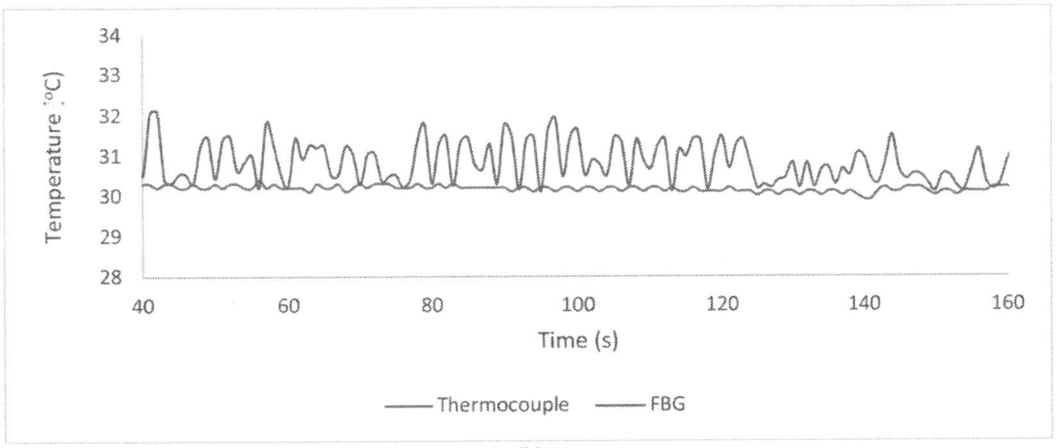

(b)

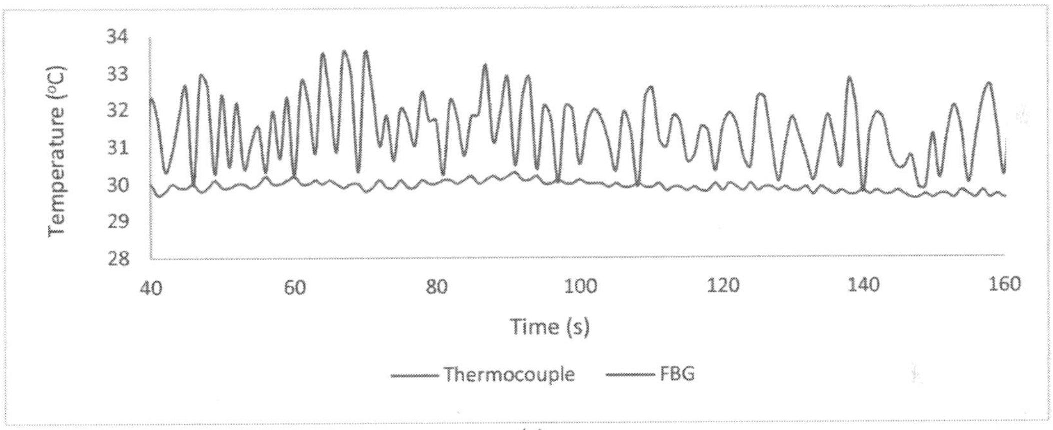

(c)

Figure 4. The waveform of the temperature response towards time for the FBG sensor and thermocouple sensor for each activity: (a) resting (b) walking (c) climbing stairs.

Table 2 gives the measurement variables of the FBG temperature FOS at resting, walking and stair climbing.

Table 2: Comparison of SNR, Peaks and maximum amplitude for resting, 5-minute walking and 5-minute stair climbing.

Activities	Signal-to-Noise Ratio (°C)	Number of peaks during 80s	Maximum amplitude (°C)
Resting	1.43	23	30.62
5-minute Walking	1.69	24	32.09
5-minute Stairs Climbing	3.23	26	33.55

From Table 2, stair climbing has the highest SNR value of 3.23°C, followed by walking and resting with Signal-to-Noise Ratio (SNR) values of 1.69°C and 1.43°C respectively. Stair climbing and walking are two aerobic exercises that require deep breathing to produce energy for the muscles during physical stress and exercise beside produces warm air during exhalation due to the post-aerobic exercise by the test subject.

4. Conclusion

In this work, an FBG temperature sensor is capable of monitoring breathing behaviour for different physical activities; resting, walking, and stair climbing. This FBG temperature sensor is used to monitor breathing behaviour and prevent abnormal physical conditions. This FBG sensor is a low-cost sensor and is light-weight, so it can be designed and fabricated into a wearable system to monitor the respiratory rate and health rate of a person.

5. Acknowledgements

Funding for this work was provided under the grants FRGS/1/2022/STG07/UIAM/02/1 and iPQC Operating Grant.

References

[1] Tavares, C., Leitão, C., Lo Presti, D., Domingues, M. F., Alberto, N., Silva, H., & Antunes, P. (2022). Respiratory and heart rate monitoring using an FBG 3D-printed wearable system. Biomedical Optics Express, 13(4), 2299.

[2] D. Presti et al., "Smart textile based on FBG sensors for breath-by breath respiratory monitoring: tests on women," undefined, 2018.

[3] Ekman, B. Kjellström, K. Falk, J. Norman, and K. Swedberg, "Impact of device-guided slow breathing on symptoms of chronic heart failure: a randomized, controlled feasibility study," Eur. J. Heart Fail., vol. 13, no. 9, pp. 1000–1005, Sep. 2011.

[4] Y. Liang, A. Mazzolini, and P. Stoddart, "Fibre Bragg grating sensor for respiratory monitoring," ResearchGate, Aug. 10, 2006.

[5] C. Mayoral, J. Gutiérrez, J. Pérez, M. Treviño, I. Velasco, P. Cruz, R. Rosas, L. Carrillo, J. Ríos, E. Apreza, and R. Laguna, "Fiber optic sensors for vital signs monitoring. A Review of its practicality in the health field," Biosensors 11(2), 58 (2021).

[6] C. Massaroni, M. Zaltieri, D. Lo Presti, A. Nicolò, D. Tosi, and E. Schena, "Fiber Bragg grating sensors for cardiorespiratory monitoring: A review," IEEE Sens. J. 21(13), 14069–14080 (2021).

[7] Mishra, V., Lohar, M., & Amphawan, A. (2016). Improvement in temperature sensitivity of FBG by coating of different materials. Optik, 127(2), 825–828.

AUTHOR INDEX